A Man of the Woods

Experiences collecting 7,000 woods of the world

A Man of the Woods

Experiences collecting 7,000 woods of the world

Richard Crow

Forest Products Society
Madison

Library of Congress Cataloging-in-Publication Data

Crow, Richard, 1936-
A man of the woods : experiences collecting 7,000 woods of the world / Richard Crow.
p. cm.

Summary: "Part travelogue and part informational wood book, Richard Crow describes how his business and hobby developed into the largest privately owned wood collection in the world"—Provided by publisher.

ISBN 978-1-892529-54-1
1. Crow, Richard, 1936—-Travel. 2. Plant collectors—Great Britain—Biography. 3. Wood—Catalogs and collections. 4. Trees—Identification. 5. Herbaria. I. Title.
QK31.C877A3 2009
674.0092—dc22
[B]

2009014934

This book has been published by the Forest Products Society (www.forestprod.org) in cooperation with the International Wood Collectors Society (www.woodcollectors.org).

ISBN 978-1-892529-54-1

Publication No. 7219

Printed in the United States of America.

09051000

Preface

In the stables of my Cornish farmhouse rests my collection of samples of 7,100 timbers of the world. It appears to be the largest privately owned collection of its kind. And this year it is 55 years old. This is the story of how it all came about. It is partly a travelogue of wood-hunting adventures all over the world, partly a tale about friendships which have lasted for years, made by corresponding worldwide, often in the most hostile of political climates, and partly about encouraging, cajoling, and beseeching total strangers to head up mountains or into the outback to collect wood for me. Mostly it is the story of my passion for wood which has its roots buried deep in my family tree and continues to flourish now.

It is my way of saying thank you to all of the people, including many of the world's postmen and women, who have helped me in countless ways. Most of all it is my expression of gratitude to my long-suffering wife who has tolerated stacks of drying wood in the house, and my frequent absences on expeditions, and for sharing the other love affair of my life.

Table of Contents

Chapter 1
Small Beginnings

"Oh, Daddy, please can I have that boat, I really would like to have that boat!"

It was a rather dejected model yacht in a little shop in the port of Dover. It had certainly seen better days. It had the comfortable but rakish lines of the 1920s and must once have looked every inch an aristocrat. Now it looked jaded and I thought it needed a new owner who ought to be me. Most of all I longed for it because it was made of wood – superbly crafted wood. But my father seemed to play it down and we wandered on to look at robust Dover Castle with its survivals from Roman times. I was just ten and the year was 1946, just one year after the Second World War. Not for the first time Dover had been the bastion to prevent invasion from France, and throughout this war it had been a "prohibited zone" for all except those who lived there, so it was quite an adventure to be there.

Not long after our visit, I was banned from my father's carpentry hobby workshop to my deep sense of chagrin, on a pretext that I had taken the edge off his chisels and banged them with a hammer instead of a mallet. The door was kept locked, and I seethed with a sense of injustice because I felt sure I would never do things like that.

Carpentry was my father's great relaxation. He was certainly no joiner, let alone a cabinet maker. That was far too meticulous for his grand strategy mind. But, give him a project like a garden shed, a hen house, or a huge fruit cage, and he would beaver away and produce a really good workmanlike product. I cannot remember ever being anything other than deeply absorbed by the projects and the lovely old-fashioned tools – auger bits, wood jack planes, spokeshaves, gouges, mortice-marking gauges, and keyhole saws. Some had been family possessions for at least a hundred years.

So I was really piqued and not a little insulted that I was not allowed into the workshop. Fortunately Christmas was approaching, so there was the prospect of some new exciting plaything to take my mind off it. At last the day came, after the ritual mutual deception of Father Christmas the night before. In the morning there were the desperate attempts to rush through breakfast and at last the frantic tearing apart of wrapping paper with my parents regulating my sister and me as if we were some over-zealous steam engines. But there seemed to be no present from them for me. Timed with consummate skill, my father and mother lifted me out of my apprehensive misery into joyous euphoria for, out of the banned workroom, came the double joy of my being allowed in again and the yacht brought in from it, now resting on a business-like stand and re-masted, re-rigged, re-painted, and re-named Astra. I was ecstatic. It was not just the joy of this lovely boat (which I still have and have recently given another refit), but it was also the pleasure of all of the workmanship with wood which was encapsulated in it. My father had outdone himself.

I certainly did not realise it at the time, but it was symbolic of the creative pains I was undergoing in the birth of my love affair with wood which has held me captive ever since and will surely hold me in its thrall until my dying day – when I'll be measured up to be surrounded by wood yet again.

Really I had no chance of resisting the charms. Not only was my father hooked, but so was his father, his grandfather, and his great grandfather who founded an embryo timber business in a little back lane in the City of London in 1795. This business was to develop into a major enterprise and last for a 165 years.

But it all began a long way away in the eastern county of Norfolk, in the hamlet of Felbrigg. In the grounds of Felbrigg Hall, which was the home of the Wyndham family (not relatives of ours!), there was a gem of a church, standing on its own and probably separated from a village abandoned in ancient time by the Black Death. In its small idyllically peaceful churchyard is the gravestone of William Harmer Crow who died in 1823. Those were the days of lovingly crafted gravestones. Under a figure of Time with his sickle and a skeleton appears this inscription:

> I hope all you that are here living
> Will pray this soul may be gone to heaven
> That from his body has took flight
> And bid all his earthly friends goodnight

William and his wife Mary had seven sons and I would like to think that the old boy said to William Miller Crow, "Look my son, in this little village there can never be work for all you young men. You must go down to London where all the streets are paved with gold. Find work there and make your way in the world."

Whether it happened like this or he rebelled and struck loose, whether he travelled down in a stagecoach or walked with a bag of a few clothes over his shoulder, I really do not know. But I do know that some time before 1795 he apprenticed himself to a carpenter and coffin maker, Robert Garnham in Benjamin Street, Clerkenwell in the City. By 1795, Robert Garnham had died and William had married Garnham's widow, thus showing his first display of business acumen. In that year, he was trading as a Carpenter and Undertaker, Timber Merchant, and Box and Packing Case Maker.

William Miller Crow's business card of 1800.

With the burgeoning success of late Georgian and early Victorian London, William clearly found that selling timber was more profitable than making coffins and other boxes. Clerkenwell was a centre of excellence in making long- case or grandfather clocks, which he was happy to service. I know from old stock books that in addition to pine from Norway, Russia, and Prussia, the stock included priceless Cuban mahogany, rosewood, and ebony. The growth of prosperity for the family continued unabated throughout Victorian times. Another William was born – William Richard – who took over from his father, and in 1864 bought the premises which had previously been rented. He then erected a brand new building suitable for a prosperous Victorian tim-

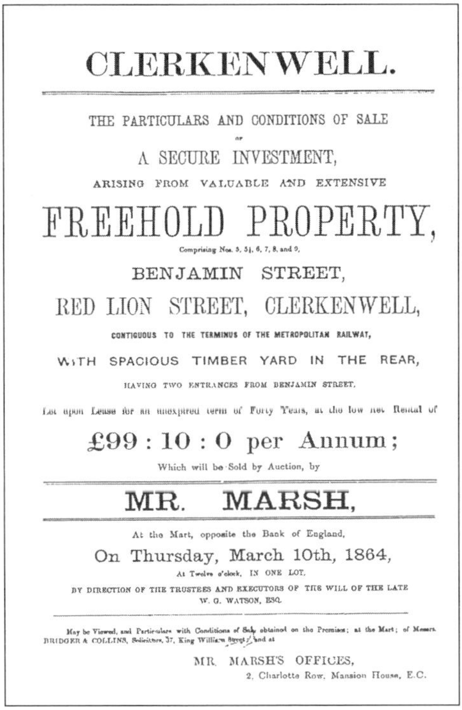

CLERKENWELL.

THE PARTICULARS AND CONDITIONS OF SALE

OF

A SECURE INVESTMENT,

ARISING FROM VALUABLE AND EXTENSIVE

FREEHOLD PROPERTY,

Comprising Nos. 5, 5½, 6, 7, 8, and 9,

BENJAMIN STREET,

RED LION STREET, CLERKENWELL,

CONTIGUOUS TO THE TERMINUS OF THE METROPOLITAN RAILWAY,

WITH SPACIOUS TIMBER YARD IN THE REAR,

HAVING TWO ENTRANCES FROM BENJAMIN STREET.

Let upon Lease for an unexpired term of Forty Years, at the low net Rental of

£99 : 10 : 0 per Annum;

Which will be Sold by Auction, by

MR. MARSH,

At the Mart, opposite the Bank of England,

On Thursday, March 10th, 1864,

At Twelve o'clock, IN ONE LOT.

BY DIRECTION OF THE TRUSTEES AND EXECUTORS OF THE WILL OF THE LATE W. G. WATSON, ESQ.

May be Viewed, and Particulars with Conditions of Sale obtained on the Premises; at the Mart; of Messrs. BRIDGER & COLLINS, Solicitors, 37, King William Street; and at

MR. MARSH'S OFFICES,

2, Charlotte Row, Mansion House, E.C.

Sale particulars of Benjamin Street. W.R. Crow & Sons, the tenants, bought the freehold at the auction.

ber merchant. It was framed out of huge square sections of pine. They were 12- by 12-inch baulks as they were called, from Memel in Prussia (which is now Klaipeda in Lithuania). They are irreplaceable nowadays.

This timber merchant's building had a saw pit where large sections of timber could be sawn into wafer-thin boards. The men used a large two-handed saw. One man stood down in the pit, presumably getting sawdust in his eyes, as he worked away at sawing the timber suspended over the pit – that job really was the pits! The other lucky man stood on top and held the other end of the saw, and together they guided the cut. It is quite extraordinary how thinly the pine boards could be cut for clock backs, or mahogany boards for table tops over two feet wide could be cut to just an inch thick, so accurately that they would fit together perfectly.

Linking the four floors of the warehouse was a shaft, so that timber or doors could be raised to or lowered from the flat roof where seasoning mahogany was stored. There was a truly Victorian office too, with high sloping mahogany desks and high stools where men marked off stock with round rulers of ebony. Even in the late 1950s when I worked there, the office was much like this, with only a telephone and a huge cumbersome calculator called a comptometer as concessions to modernity.

In Victorian times, William Richard Crow enjoyed his prosperity and lived very comfortably in Croydon, just south of London, having been born in one of the humble buildings in Benjamin Street. When he died in 1888, the inventory of his house listed many things including an enigmatic stuffed dog and over 600 bottles of wine. I know that he enjoyed many a good tipple, which shocked his rather precious son, William Merritt Crow, who was next in the line. "Don't let little Willy see," he would say as he downed a little more of what he fancied, which did him good.

Shocked or not, William Merritt Crow was a good businessman and expansion continued, bursting out of the confines of Benjamin Street. In the early 1900s, the company had an active office in Cannon Street, in the commercial centre of London, from which vast quantities of timber were bought and sold without ever physically passing through Crow hands. Then a wharf was bought in Greenwich a little further down the River Thames, into which timber, imported from Scandinavia and Canada into the bustling Surrey Commercial Docks, was transferred by barges and landed with every piece being manhandled. In our highly mechanised age of packages, containers, cranes, forklifts, and huge articulated trucks, it is difficult to imagine the scene then, with horses and carts, a jungle of trestles and running boards, and an army of men called deal porters with thick

William Merritt Crow with his wife and family, 1902.
My father is standing at the back on the right.

leather shoulder pads carrying the timber from the barges into tall man-made stacks in sheds. But carry it they did, and I can well remember seeing it being done by cheery cockneys of indomitable spirit.

William Merritt Crow, my grandfather, was a very different person from his father, and from me. From his early days, he displayed a keen zeal for non-conformist religion. I am no expert on the Plymouth Brethren, but apparently it all began in Plymouth in 1830 among old India hands and spread like wildfire. It was observed all over the country, and it spread to Dublin, the Continent, and the Empire, with, for some reason which may have been my family's zeal, a strong following in the timber trade. The basis seems to have been a protest against hierarchical religion and a conviction that only its select band of adherents could inherit eternal life. "Be not unequally yoked with unbelievers" and "wherefore come out from among them and be ye separate and touch not the unclean," it says in the Bible. These were their tenets, and contact with others had to be reduced to the minimum to escape risk of contamination.

I do not know how my grandfather's very suitable temperament came into contact with this movement or how he met his future wife, Georgina Acheson from Dublin, one of the most devout of Brethren. Her family was Scottish, but it had also established itself in Dublin and in Northern Ireland where it had as its head the Earl of Gosforth.

But gone were the days of wine and roses of his father. A period of austerity set into the family business and a Dickensian but well-intentioned regime came in, exemplified by gifts of Bibles as Christmas presents to staff.

Into this austere aura, my father, William Roberts Crow, was born in 1878, followed by his three brothers and two sisters. Safely cocooned in the certainty of divine favour, there was little opportunity for contact with other philosophies until my father, who had been to a good Plymouth Brethren school near the city of Bristol about 100 miles from home, had the opportunity to go to Oxford University. Such a dangerous exposure was not acceptable and was vetoed. As a compromise, however, he could go to London University provided he returned home each night. He saw this through, but even this tight regime did not protect him from the outside world and that was the end of any allegiance to the Brethren that my father might have had. It took great courage to break out of the straight-jacket and establish contact with the damned, and drive a wedge between him and his parents, brothers, and sisters. But this he did, and my father would never have been comfortable in that kind of environment. He was bulky, genial, and largely non-religious, and shared many of the characteristics of his grandfather whose lives only overlapped by ten years.

Jack Crow,
English scholar and eccentric.

My father, "Bertie," and my uncle, Alec, from their separate sides of the great religious divide, joined the business, and in 1904 my father married for the first time. Two sons, Jack and Basil, and a daughter Betty, followed soon after. Jack perhaps followed his father's ambition when he did indeed go to Oxford. Jack graduated in English and went on to become an authority on Shakespeare, co-writing this subject in the *Encyclopaedia Britannica* of the time. Incredibly, for a time he was also a New York-based boxing correspondent for a London newspaper. To us, Jack was a mystical, almost mythical, jovial and rotund figure held together by one strained button of his jacket, who seemed to alternate between lecturing at London University and Stanford in California, and then being Professor of English at Pittsburgh University. He certainly took to things American but not without a few mishaps, as this excerpt from a 1930s letter from New York to his father Bertie shows.

Dear Boit,

You soitainly desoive a letter. I have been meaning and meaning to write and always waiting for some real piece of news preferably good. At last I have good and bad rolled into one.

I have found the last three Mondays inauspicious. On July 6, I played a strenuous game of handball and removed acres of skin from my right foot (loose sandal). On July 13, I went one better than this. I was going up the heavily carpeted stairs in a movie theater to the Gents' Comfort Station (still makes me laugh) when I tripped. I crashed my chin down and it split open. I had to have two stitches in it. They are out now, but it isn't absolutely healed. It bleeds timidly after a shave.

That was nothing to the exploit of July 20 which consisted of stepping into an apparently non-existent subway train. I thought I was stepping into the train from the platform, but there was a believe-it-or-not 14-inch hiatus, the platform being steeply curved, and I went into the abyss.

I had to be lifted out and helped into the train (fortunately motionless) by passengers. I thought I was only stiff and shocked but I had about 12 square inches of abrasion on the right shin. I had subsequently to be x-rayed for suspected greenstick fracture of the fibula. Also further visits for dressings.

The result (cheerful) is that I am claiming some pleasant damages from the Subway Co. Silver Lining. I still limp and am still under treatment but much better...

And that is about all. I hope that everything is lovely in the garden etc., all members of the family flourishing etc. etc.

Much love to all.

Back in England, his base was a tiny apartment in Denmark Hill in London, in which lurked a bed which could eventually be found amid piles of books, many of them first editions which he had been given to review for the *Times* Literary Supplement. It was always a happy event when he came to see us much later in life with a haversack containing unspeakable things, from which he always produced presents for us. He smoked a revolting pipe and cut his own tobacco from a plug of hard block substance. Like an Australian aborigine, he would go walkabout in England and on the Continent. Once, when he was sleeping under a culvert, he was woken by a policeman and in reply to the "Allo, Allo, what are you doing here?," Jack said in his deep voice, "I'm a burglar." Uncertain about this, the bobby

Unloading timber from a barge at the wharf at Greenwich in the 1930s. Alec Crow looks on.

led him to the station where, plied with copious cups of tea, he kept them entertained until the morning when he was released.

Jack was not remotely interested in the family business, and Betty married one of the Rowntree chocolate family, but for Basil the call was irresistible. So, between the wars my grandfather presided with religious fervour over the family estate, aided by his two sons and Basil.

The depression of the 1930s affected business of all kinds and presented a very different environment for the company after the vibrant prosperity of the Victorian era. The seemingly endless demand for timber tailed right off. My grandfather listed several major timber companies of the time which had collapsed. Our company had to borrow heavily to pay for all of the imported timber it had purchased but could not sell. Fortunately, they were able to pay back the debts and survive, which must have called for much skill and religious determination.

In 1933, my grandfather died, leaving my father and uncle to run the business. My father had already suffered badly; his wife had been ill for several years and died in 1929. One year later he married his secretary, Beatrice, who was to become my mother. My sister, Anne, was born in 1933, and I drew my first breath in 1936. The family was blissfully happy and the business had survived, but before long new problems loomed menacingly.

Chapter 2
Aspects of War

At the start of the Second World War, my father and mother bought a "nice" house in the town of Sevenoaks, south-east of London. Benjamin Street continued as ever, and for some time they retained the wharf at Greenwich. The bombs fell all around it, however, and eventually a map was found in a shot-down German bomber which highlighted the wharf with its timber as one of the good places for starting fires – so they sold it and retreated to a small office in Sevenoaks to wait out the war. My father and uncle were too old even to serve in the Home Guard, or Dad's Army as it was called, but Basil joined the Intelligence Corps and my uncle Alec's son Edmund, of whom we were all fond, was the second in the family to break away from the Brethren (distancing himself from his immediate family just as my father had). He signed on as aircrew in the Royal Air Force and became a navigator in Stirling bombers.

The day war broke out, I was only three, but I was nine when it ended. I have very clear memories of much of that time. Who could fail to, living in and among such an experience? The one thing I do not remember is fear. Whether this was because our parents were so good at concealing their fear, or whether my sister, Anne, and I were too young to understand the dangers, or whether fear was unthinkable I do not know. All of my memories are of excitement and the thrill of the unexpected.

Our nice house in Sevenoaks, which was on a "nice" road where every garden was at least three-quarters of an acre, had a tennis court levelled out of the slightly sloping ground. My father, in one of his bursts of enthusiastic energy and patriotism, dug up half the tennis court to grow vegetables for the Dig for Victory campaign and situated a dug-out air-raid shelter at one corner, made of railway sleepers or cross-ties for the walls, flat roof and subterranean floor. Further up the garden he built a huge fruit cage and added a chicken house, so we were well prepared.

In the house, my mother made air raid curtains out of black material, and in the dug-out were bunk beds, food, cooking utensils, and everything needed for living. I remember Anne slept on the bunk above mine and laughingly dropped sporadic canister bombs of Sanatogen tins onto me. In spite of all of the preparations, my father insisted on sleeping in the house on most nights, with his loaded First World War revolver beside him. Most nights, the Air Raid Warden would look in to see if we were all right, too.

As we were in Bomb Alley on the German route into London, several houses disappeared around us and we never knew what to expect in the morning when we left our dug-out. Sometimes windows would be broken. Other times there would be a different kind of "window," which was the code name for silver foil which was dropped by the Germans and landed all around us. It was used to jam the radar, although we didn't know its purpose then.

Despite the warnings, we frequently explored bomb craters and picked up lumps of shrapnel. Nothing seemed to deter us little boys, not even the butterfly bombs and other devices intended to blow up little boys. One boy even had a pencil case made from a mine casing which he regularly took to school. To the New Beacon School we walked the mile each morning with our gas masks strapped over our shoulders. All around us were barrage balloons floating at the top of long leads to intercept aircraft. There was one of these in the large school grounds. Just a mile away, concealed in dense woods, was an enormous ammunition dump which, if it had gone up, would have been very interesting.

The New Beacon School of that time would be an enormous anachronism now, but in the war it was perfect. It was run by two brothers who, having fought in the First World War, decided to run the school as if it was an army barracks. Major Frank and Captain Cecil Norman presided over a school where there not "houses" but companies, Drake, Nelson, Marlborough, and Wellington, with Company Leaders and Army versus Navy rugby matches. Boys were known by their surnames only. Standards were high and discipline strict. Punishment was at least mildly painful and at most corporal. A nasty rap on the knuckles with the edge of a ruler was sufficient to keep most boys in order but six of the best was a fearsome ultimate sanction. Tables were recited in methodical fashion and homework, or "prep" as it was called, was seldom neglected. Sport was de rigueur and taken very seriously, although I wasn't very good at it.

It all sounds terrible when recounted like this, but I mostly have happy and proud memories of the school resolutely carrying on as normal in such difficult times. I say "mostly" because there were unpleasantries such as the same food on the same day each week, some of which was execrable. I shall never forget Thursday's stew which was said to be neck of mutton but which we believed contained all sorts of revolting animal parts.

Another part of the formative years of my love of wood took place at the school. This was the optional carpentry lessons. They were run in a ramshackle wood building by a Mr. Clarke who had been Ship's Carpenter on square riggers. Highly structured, the lessons taught us to make boxes, do dovetail joints, and use sandpaper ad infinitum. Woe betide any small boy who forgot to attend his optional lesson. Mr. Clarke would gatecrash the class the boy should have opted out of attending. In his hand, he had a short length of highly skilled knotted rope, which he turned in circles menacingly and said, "I'm on the warpath." This never happened to me.

Back home the night bombing went on and the dog fights took place over our heads. Toward the end of the war, the doodle bugs, the V-1 pilotless bombers, came droning over. Suddenly their engines would stop as they ran out of fuel and you would count up to ten before there was an almighty bang as they hit a target some ways away. We knew from experience that if they stopped overhead you were perfectly safe, but our next-door neighbours, who had recently arrived from South Africa, used to scurry terrified to their air raid shelter. I am afraid we had become so blasé that we used to stand on the remaining part of the tennis court and roar with laughter at their discomfiture.

While all of this was going on, Basil was in the army and could be contacted, but we heard one day that Edmund was missing after a bombing run to Germany. Several weeks passed before we learned that he had been shot down and mercifully was now in a Stalag Luft prison. When at last the war was over, Edmund came back as a most charming and relaxed, debonair man. Gone were all traces of his Brethren upbringing. Instead he bought a racy vintage Bentley car and adopted a style of fast living and copious drinking which we adolescents greatly admired. He was our hero. He told us gripping tales of his experiences as a prisoner-of-war. Near the compound where he was, was a compound for Russians. The Russians were troublesome, getting out at night and going on the rampage, so the Germans let Alsatian dogs loose to deter them. Unfortunately for the Germans, and the Alsatians, the Russians had eaten them by the morning – the Alsatians that is.

Meanwhile, back at the family business, seeing the end of the war in sight, my father and my uncle had decided to buy a derelict wharf at Belvedere on the southern side of the Thames, down river from London and opposite the Ford Motor Company at Dagenham. It had a frontage on the river and looked like a superb place to set up their post-war enterprise. Even this site had not escaped the bombing, however, and there was the wreckage of a Candle Works on one part of the 26 acre site. Pleased as punch, I am sure, at the prospect of developing this site, my father and uncle set up shop in some of the wood buildings still remaining and, monarchs of all they surveyed, looked out over a scene of total desolation.

One day as Uncle Alec was taking a constitutional walk around the run-down estate, he found a man setting up levels and theodolites (or whatever surveyors set up) and peering intensely through them. Intrigued and no doubt feeling testily territorial, my uncle asked the classic landowner's question of an intruder, "Can I help you?" But this enquiry was met with a disconcerting, "No, I'm all right thank you." "Well, what are you doing here?" "I'm afraid I cannot tell you." "But I am the owner of this land." Still nothing was forthcoming. I do not know how this impasse was broken but eventually the family was sworn to secrecy – the War Department intended to build three Mulberry concrete floating harbours in dry docks that they would create, then breach the bank of the Thames and float them away across the Channel for 'D' Day, the invasion of Europe. How my sister

Crow's Wharf at Belvedere in Kent, fronting the Thames, 1960.
It was purchased as a derelict site in 1944 and soon after three Mulberry concrete floating harbours were built by the Government in docks hollowed out of the site. The banks were breached and the harbours floated off to France for D-Day.

and I, excited youngsters, managed to keep this secret I cannot imagine, especially as we were allowed to go up to "The Wharf" and see the huge holes that had been dug in the ground. Nearly filling these up were the shells of the harbours. Work went on 24 hours a day, and I can clearly remember the make-shift canteen where the exhausted mud- spattered workers unwound before heading back to the hell holes.

Not long afterwards, two things happened. Our nice quiet road in Sevenoaks, where traffic was an event, was suddenly filled with army lorries and friendly soldiers. This was great fun for us kids as nothing ever happened like this. The next day they disappeared and we heard that the Thames banks had been breached and the Mulberry Harbours floated away.

All that now remained was some rather unpleasant holes where the site of the new Crow world should have been. The Government showed reticence similar to the first surveyor about what was going to be done about that. Britain's war was won, however, and we were finally offered financial compensation to restore the site. This was wonderful in theory, but in practice no amount of money could provide the equipment to push all of the debris back into the holes. Mechanical equipment was at a premium and the family investment now looked like an abandoned, barren gold mine. Not short on tenacity, my father and uncle fought and fought to have the site reinstated by the Government; only after a decision in the House of Lords was it agreed that it would all be leveled back for them. This was another time of great excitement for us little boys on our school holidays. My father would take me and some friends up to the wharf so that we could watch the spoil going back with as much impetus as when it was all taken out. Instead of what we now call diggers, there were the much more genteel excavators which were cranes with buckets and other novel tools with good Anglo-Saxon names like bulldozers and dumpers. The dumpers were driven by Italian prisoners-of-war. They kept getting stuck in the mud and had to be dozed out by the bulldozers. All this sort of thing we take for granted now, but in the mid-1940s it was literally an earth-moving experience for us impressionable boys (and secretly I am sure for our elders who naturally appeared to take it all in stride).

All of this, of course, happened just after the war was over, but I need to go back to fill in a few more earlier details, back to being at home when the doodle-bugs were droning and running out of fuel overhead and the war was still a reality. The most exciting bit was to see the Spitfires and Hurricanes catching up with the V-1 doodle-bugs and tipping their wings over to one side to deflect them away to somewhere safe. The doodle-bugs were very visible and audible, and therefore menacing, but they were far less dangerous than the V-2s which followed. Nobody could see or hear the V-2s, we were told, because they flew so fast and so quietly, but they could inflict much greater damage. We never did see one, so I don't think we really believed in them.

Fortunately their reign of terror was short, for we were winning the war and even as children we could sense that atmosphere. My father who seemed to have thrived on the First World War was, I am sure, a frustrated fighter in this Second War. He listened avidly on the wireless to the news, following every reported advance on maps, by moving pins further forward. His exhilaration was infectious as unpronounceable names of distant towns were bandied about in conversation until they sank back into obscurity as the war wagon moved on.

At last one day the war was over. It seems silly to say it now but the feeling I remember most is one of anti-climax. There had always been a war and this was part of living. We were never going to lose, in fact we were certainly going to win. So what were we going to do now with no dug-out to go to and no friendly barrage balloons to fill up the sky? I suppose we were so well protected and motivated as children that we did not take in the awful realities of war, and it was only when it was all over and we were growing up, and the truth about the terrible atrocities began to emerge, that we began to take a more sanguine view. But I do remember the short-lived yet tremendous euphoria of the celebrations before everyone knuckled down to the harsh realities of rebuilding Britain.

Chapter 3
A Little Learning

In 1949 at age 13, I moved on from the New Beacon School to nearby Tonbridge School. Founded in 1553 by London benefactor Sir Andrew Judde, it had originally been a charitable school funded by his property interests in London. For about 300 of its 400 years or so, it had been an old-fashioned grammar school, but it seems that in Victorian times it burst out of its acorn shell and decided it was going to become one of the biggest trees in the forest and that seems to get better every year.

It was just as well that the New Beacon school had a fairly tough regime because Tonbridge had plenty of that, too. The traditional structure of these schools was very well established, the rules were clear, and the hierarchy was unassailable. (It has changed a lot since then.)

The means of breaking down the total 500 pupils into manageable units was the house system. There were eight of these for boarders and two for day-boys. The boarding houses really were houses where 50 boys, with ages spread from 13 to 18, lived as a residential unit under the tutelage of a housemaster and his wife, and usually a much-admired matron. There was something intensely monastic about this house system. Novitiates (novi) had their communal room (the prep room), while junior inmates shared their cells (studies) with one other. Senior brothers had a solitary cell, and the abbot, allowed special dispensation to be married, ruled from his monastic office (the study). The abbey kitchens dispensed wickedly luxurious food conjured from meager resources of ration books. All of the inmates loved the matron and longed for the daughter of the housemaster to reach maturity so that they could dream Carmina Burana dreams of her.

Of course when it came to teaching, the regime was completely different. Boys of the same age-range from each of the houses were in forms learning anything from classical Greek through algebra, geometry, and trigonometry to modern languages – and again the option to learn rather more advanced carpentry. The standards were excellent, and it was inspiring and stimulating to be learning among some talented people.

~ ~ ~ ~ ~ ~ ~ ~ ~ ~ ~ ~

In the school holidays we had some superb holidays in England, but we had never ventured abroad as a family. In fact, "abroad" just after the war was not an easy place to reach. But when my sister was 18 and I was 15 and still at Tonbridge School, our parents decided that 1951 was going to be the year when we tackled "abroad" as a family. But complications arose.

My father, having married twice, was really old enough to have been my grandfather. I was born when he was 56, so I was in my mid-teens when he was in his early 70s and his health began to deteriorate. Before he had become ill, my parents had arranged a two-week holiday to Zell-am-See, a lovely little town by a lake in Austria. We were to travel by train from London to Dover, then cross the channel by ferry, then on by train from Boulogne to Innsbruck via Basel in Switzerland.

With our father's illness, we faced the option of canceling everything or the courageous choice of Anne and me going on our own. Our mother, always a supporter of every good adventure, took little time to decide that Anne and I should go.

Beside ourselves with excitement and blissfully ignorant of what to expect, at last on the appointed day, we boarded that train in London, crossed the channel, and arrived at Boulogne which, only six years after the war, was still looking badly damaged.

It was evening as we boarded the French train and headed off into the night. Although someone befriended us somewhat, we really were innocents abroad as we sedately crossed France in the dark. At last in the morning we reached Basel, where we knew we had to change trains for Innsbruck. We thought they said, "Wait for the next train on this platform." Unfortunately they said or should have said, "The second next train on this platform." But, coolly and confidently, we climbed aboard the first train which, shortly after, pulled out of the station. We were thoroughly enjoying the mountain scenery, so it took some time for us to realise that none of the other tour people were on the train. Our French was no credit to our excellent education and our German was non-existent. Everyone else had that satisfied look of knowing exactly where they were going, so we adopted an air of external calm, while inside we were experiencing our first ever feelings of mild panic at being in a foreign country, lost, with no one understanding what our problem was.

All Central Europe suddenly seemed to be a menacing, nasty place where they had vicious wars. I imagined ending up in Berlin or Krakow or Moscow and being locked up for illegal entry (six years before, we would have been). The train ambled on, never stopping, and, nice although the scenery was, each mile seemed to be taking us further into oblivion. At last it stopped at a place called Buchs (indelibly recorded in my brain). We clambered out and the train moved off. "Innsbruck?" we said hopefully. "Ja, next train," they seemed to say but, our confidence shaken, we had a long apprehensive delay. At length another train pulled in and there were the other tourists, jubilant at seeing us there much chastened. Thankfully, we climbed aboard.

We had a wonderful holiday in Zell-am-See. We climbed mountains, we crossed glaciers, we went on coach tours, Fritz fell in love with Anne and wrote to her for months. In the evening there was the "Tiroler Abend" with thigh-slapping, concertina-playing, leder-hosened Tiroleans singing and dancing; we thoroughly enjoyed it.

Only one other incident heightened the drama on that holiday. We had taken a coach tour across a little part of Germany to see Berchtesgarden, Hitler's old mountain haunt. As we crossed into Germany, our numbers were counted. While we were over the border, two tourists who had visas left the coach. When we wanted to cross back into Austria, our numbers were again counted and we were found wanting. What to do? Well, go back and find the two missing people. Despairing of any chance of seeing them again, we just happened to see them outside a shop in a little town. We bundled them aboard and told them that, visa or no visa, they were coming back with us!

This was a wonderful hors d'oeuvre for the rigours of foreign travel in which I would partake later.

Chapter 4
Swedish Rhapsody

Unfortunately, my father's health did not improve. I think his system was just wearing out, and, in November 1952, it wore out altogether. Despite the obvious warnings, it shocked us dreadfully, especially my mother who had been very happy to play second fiddle in the marriage. She was over 20 years younger than he and had been his secretary. But she had great reserves of courage and resourcefulness and was a very well organised person. She soon filled the breach, while my sister and I plodded on at school.

By now I was quite determined that I wanted to continue in the family tradition and be one of the joint inheritors of the family estate. My Uncle Alec had been alone at the head and Edmund, his son, now joined him. My half-brother Basil had decided to emigrate to Canada so there seemed plenty of room for me. Lacking the fatherly guidance I suppose I really needed, I decided that I would just get "0" levels (ordinary level grades in exams) and then call education a day, as the only thing I wanted to know about was wood. Fortunately, my housemaster took me in hand and persuaded me to stay on at Tonbridge to get two "A" (advanced) levels, although I could not be persuaded to go on to university (my silly decision).

Next on the menu was National Service, our compulsory draft, which loomed with an air of inevitability. Three things were now bugging me, which have bugged me ever since – the love of wood, the yearning for travel, and the magnetic attraction of the family business. National Service was going to get in the way, but I would not be called up before March 1954 when I was to be 18, and it was now only August 1953.

Canada seemed to be the ideal place to go for the six months. I had heard of young timber people like me hopping on a boat at Liverpool and working their passage across to Quebec or Montreal. Then onward they went on the romantic train journey that took several days to cross the prairies and climb into the Rockies at Banff, ending up at Vancouver. There they ended up among the tall trees in a logging camp with the lumberjacks, and built up their muscles. This was the nearest thing to backpacking or a "gapyear" of the time, and I decided this was for me. I contacted one of the shipping companies who, under a little pressure, overcame a reluctance to cooperate. They would take me not to Quebec but to Boston, and from there I could still get on course to Vancouver.

My father had corresponded all through the war with a very good friend who was a lumber man in Toronto, Jim Terry. He had kept Jim enthralled with heroic tales of war-torn Britain, which had been published in a Toronto newspaper. In return, Jim had sent us gorgeous food parcels which usually got through to us. He had contacts in Vancouver, so I decided to write to him. He said it could be done, but he wouldn't recommend it because the lumber trade was going through a bad patch (I can hardly remember a time when it wasn't). I would have to rough it with some pretty tough people, but, if I insisted, he would give me a contact.

By now I was beginning to get quite excited about this adventure. When I had a positive letter from the Hillcrest Lumber Company (which sadly is no more), the only thing left to do was get a work permit. But here I ran into a problem which was to prove fatal to the plans. As I was not quite 18, I would have to be met at the station of final destination (after all that traveling) by a responsible adult who would vouch for my safekeeping. Understandably they said at Hillcrest, "Hell, if he can't look after himself at the station he can't fend for himself here." To wait until I actually was 18 and make the plans, only to be called up, was far too risky, so with deep disappointment I had to abandon the idea. But then almost immediately an alternative cropped up, perhaps not quite so exciting or romantic but an alternative. My Uncle Alec asked me if I would like to spend six months working in a sawmill in Sweden. Of course Sweden has supplied Britain with timber for years, but I had not really thought of Sweden as a country I was yearning visit. As I got into the details, however, I became increasingly enthusiastic. When I finally boarded the Swedish Lloyd ship "Saga" at Tilbury with only some clothes and a minute Swedish/English dictionary, I was raring to go.

On the overnight trip I learned that I am not a good sailor, and when I left the ship at Gothenburg the first Swedish word I groaned, garnered from my dictionary, was toalett. I now had to find a train to the little backwoods town of Hällefors, which is almost exactly between Oslo and Stockholm, a little north of Lake Vänern. It is surrounded by forests and there are several picturesque lakes. But I had had no time to prepare myself for anything Swedish, so I knew nothing about that.

The Sweden of 50 years ago was very different from today. Sweden, of course, remained neutral during the war, but it had not totally escaped privations. Certainly the work ethic in the country was very rigorous, and union strength was not much developed in Hällefors at least; however, social conditions and living standards seemed quite well developed. Hardly anyone spoke English but several were keen to try.

In Gothenburg, I managed to find my way onto a train which moved smoothly through northern pine and spruce forests. Eventually it reached Hällefors, where I was relieved to find someone there to meet me. I was to live in one room of a small house of a painter and decorator who spoke not a word of English either when I arrived or six months later when I left. This did not matter too much as necessity forced me to learn a fair amount of basic Swedish to survive.

First I had to hire a bicycle. It was one of those heavy, old-fashioned ones with a brake which worked when pedaled backwards. Now I had a good means of getting to work. The town had grown from silver and iron mining and working by several separate enterprises in the 1700s. In 1864, Hällefors Works Company had been founded to amalgamate these operations into one. With the growth of sawmilling from the ample surrounding forests, it was now partly a steel company, partly timber, and partly paper. Most of the men at Hällefors worked for the company.

All I knew was that I had to start work at 6:00 in the morning and I would finish at 4:00 in the afternoon unless there was overtime until 6:00. This was not optional, and I would be told at lunchtime whether there was overtime or not that day. Breakfast would be at 8:00 for 20 minutes, and lunch was at noon for 40 minutes.

This was my very first job and I was enthusiastic, if apprehensive, to start. I had no idea whether these were typical hours for a job and I came to believe the pace was quite normal – but it was killing!

On my very first morning of my very first job, I rode in early to the sawmill ready for 6:00 and ready to meet the manager, Herr Murren (pronounced Muir-rain). Herr Murren, I soon discovered, was an absolute brute of a man, a huge stocky Laplander. Everyone was afraid of him. I thought he was like a rampant bull, but they called him Elefanten.

The mill was a huge rambling range of timber-framed buildings by the riverside. I expect it has long ago been pulled down, but it should have been kept as an ancient monument. Logs felled up-river were floated down in their thousands to the log storage area. Logs felled downstream were brought up by rail and tipped into the river. All of them were lifted out of

Hällefors sawmill, Sweden, 1953. In the foreground is the grading table on which timber would pour out of the mill and end up in a heap on the ground if things got out of hand.

the water and passed through an incredibly noisy grinding machine which stripped off all of the bark before either being added to a huge pile of landed logs or being returned to the river ready for sawing. Stacking the logs on land was one of the most dangerous jobs at the mill as, newly de-barked, they were very slimy and slippery. The men whose job it was to do this had very sharp steel teeth fitted under the soles of their boots which were sharpened every week.

The logs in the water were channeled according to their diameters, ready to go into the mill. There would then be a long run of product of one size of one species, furu (pine) or gran (spruce). One-by-one the logs would go up a ramp into the bedlam of the mill. Inside was one of the most frightening scenes I have ever seen or heard. The noise was phenomenal and huge hunks of timber seemed to be flying all over the place.

The star turn was the frame saw of which there were four, so tall that they occupied two storeys of the mill. The lower part consisted of a huge fly wheel which drove a vertical piston. On the floor above, the piston took the form of a frame in which a number of saws were fixed. The whole thing oscillated up and down at great speed, and, as the log was fed into the frame, several saw cuts were made at once. At the centre, there was a section with two opposite faces sawn parallel up to 11 inches apart, and the other opposite faces were still rounded. On the edges of this hunk of timber, there were a number of boards about an inch thick. These shot off at great speed on conveyors to adjustable parallel circular saws which trimmed off the rounded or "waney" edges. The main hunk of timber was then turned on its side and went through another frame saw which converted the one piece with rounded edges into two pieces without rounded edges. These pieces then came at a manic speed out of the mill into broad daylight.

This was where my job and Björn's came in. These pieces had to be graded into three different grades and manoeuvred into appropriate railway wagons from an elevated position. The trouble was that, regardless of any difficulties that you might be having in this manoeuvring, the pieces of timber still kept coming. It was all right if the sizes were small, such as 4 by 2, but the biggest, 11 by 3 and up to 20 feet long, were very heavy and, if you got yourself into difficulties, they would pile up or slide off the end of the elevated platform in a huge jumbled heap. There was no communication with the inside mill to say "help." I soon learned that if everything was under control, you used leverage to "make the timber work for you," but if it was out of control you just had to let the pieces run on to the ground and pray for a breakdown. "Stanna" meant stop, that blessed hiatus when there was temporary trouble at the mill and you could rush around picking up the pieces with no leverage to help you. "Slut" was an even more blessed word because that meant the finish of the run and a stop for breakfast, lunch, or for the day. At the first "slut" at 8:00 on my first day, I had done just two hours work but I thought I had done two days. By dexterous use of the dictionary, I learned that the thing to do now was cycle around to the works canteen, the Bruksbaren, for 20 minutes basic breakfast.

Juli-i-Kurva cycling to work at Hällefors sawmill. Trains had the right of way.

In the immaculate canteen, gleaming with Swedish clinical efficiency and contrasting with the dusty shambles of the mill, I found that the thing to do was buy a card giving five working days of meals, and tear one piece off each time. I was learning.

There were a number of characters in the mill apart from the brutish Elefanten. There was Juli-i-Kurva, who lived by the bend of the railway about a mile away and cycled nonchalantly to work on a strange tricycle which ran along the railway lines. He had to lift it off if he heard a train coming. There was Hans, a bluff amiable German from Hamburg who had managed to escape to Sweden and dodge the war. He spoke Swedish with a harsh Teutonic accent which even I could recognise as comic. Unfortunately, he had some years earlier put his hand rather carelessly on a circular saw and was left with only half a finger and his thumb. Undeterred, he filled in his football pools coupon weekly, holding it in what was left, while telling us tall stories about his sexual exploits in Hamburg. And there was Björn, my fellow-sufferer at the end of the production line, who pulled my leg unmercifully and caused hoots of laughter at my valiant attempts to get my tongue around Swedish words. His attempts at English words were often hilarious too, but it seemed such a pity that there was not an English audience to roar derisively at his mangling of English words.

A fire in that tinder box of a sawmill would have burnt it down in no time and put everyone out of a job, so smoking was definitely förbjuden, and instead everyone took snuff. Not sniffing it, they chewed it; and, despite notices everywhere, "Spotta ej på golvet," do not spit on the floor, that is exactly what they did with great relish, leaving a pattern of trodden brown stains.

I was rarely called Rikhard in the mill; more often I was Engelsman (Englishman), especially if there was some unpleasant or menial task to do. I had only been there a few days when Engelsman was summoned by Herr Murren into the bowels of the mill where there was an ever-growing pile of sawdust. Nowadays sawmills have sophisticated suction extraction plants, but Hällefors mill had an antiquated system like a dredger's. There was an endless chain with spaced strips of wood which rotated slowly, dragging the dust away. It had broken down. So Engelsman's job was to keep shoveling away for the rest of the week until it could be repaired. If I thought the end of the production line was hard, I soon realised that there could be much worse things than that. I fought single-handed for days as the pile grew and grew until it filled up enough space to fill a house. "Never mind," they said, "it will be repaired at the weekend," and then I had to shovel it all back onto the conveyor again.

As well as at work I was making a nice circle of friends outside. Someone had told me that Karins Matsaler (Karin's Foodrooms) was a much nicer place to eat than the canteen at work, and I could get a seven-day coupon at a good price. It was wonderful, and I imagine rather like an inn in Chaucer's time. Karin was capacious, warm-hearted, and mildly bawdy. The atmosphere was just that sort of warm one which develops around a huge roaring open fire. The food was tremendous and you were encouraged to eat as much as you could. It was here that I first discovered "sild" or pickled herring, for which I have had a consuming passion ever since.

My closest friends were Willard and Alan, neither of whom, as it happened, worked in the sawmill. They were both a little older than I was, and Alan had a Ford Zephyr car of which he was tremendously proud.

Notable about Sweden was its pretty strict ideas about alcohol at that time. Alcohol could only be consumed in public with a meal; to buy it to take home, you had to use a ration book which was only issued to you if you were 21 or married. The ration was one litre of brännvin (Schnapps) per month. My friends and I decided this was nowhere near enough, so we undertook a little weekend journey on behalf of the village, across the border to Oslo in Norway, where I learned that northern people can laugh just as much as southern. Our return was the only time I have been engaged in smuggling. The back of the car was visibly weighed down suspiciously with contraband brännvin filling the boot (trunk). When we were stopped at the Swedish customs and asked if we had any alcohol on board, our look of horror at such an idea carried us through.

I was having a marvelous time outside work, and I was even asked to come along as Exhibit A at trade union English lessons. I found it fascinating to have my language scrutinised in such detail. "Why do you begin so many sentences with, "Well, ..?" "How do you distinguish between "th" sounds as in forth, from "th" as in the?" There is neither sound in Swedish.

Even at work things were going very well, and I had been promoted to the final grading and dispatch at the timber yard. But this was rather repetitive work and it was outside. It was beginning to get very cold, and even when the sun shone it was so low that it never appeared above the timber stacks. So I decided to cut the six months back to five.

For my departure, there was a tremendous send-off. The company gave me a fine book all about itself and a reference which I have to this day. The trades union class gave me a superb Swedish glass vase. Willard and Alan gave me some bottles of brännvin, and Karins Matsaler gave me a terrific farewell party and a huge bottle of pickled herring. It had been a very good time.

Chapter 5
Collecting My Thoughts

Despite the mundane work in Sweden, I had learned quite a lot about Swedish pine and spruce (timbermen call them redwood and whitewood). Being conifers these were softwoods, but about hardwoods I knew nearly nothing at all.

At this point perhaps I should clarify what softwoods and hardwoods are. The names are generally satisfactory, but they imply a simple graduation from very hard to very soft which is not the basis of it. The distinction is a fundamental botanical one. Without going into many technicalities, softwoods are the product of coniferous trees and hardwoods are the product of nearly all other trees.

Conifers have a much older origin than hardwood trees. Their seeds are naked and borne in cones (even juniper and yew berries are modified cones). Their wood structure is more simple and primitive. The trees grow very straight like poles, with a limited number of species covering vast areas of the northern regions. Toward the tropics they fade into obscurity, and other conifer species of lesser importance occur in a band across the temperate parts of the southern hemisphere.

Because the trees of the northern forests grow so conveniently straight in such vast numbers of such few species, and because the woods are so conveniently soft and easy to work, they are the softwoods – the woods we use for general carpentry.

As the conifers decline toward the tropics, the hardwoods take over. They are much "younger" and their wood structure is more complex than softwoods. While there are about 600 different softwood species, there are probably at least 40,000 different hardwood tree species. The fruits of these trees are anything but cones (except a few), and are such as apples, acorns, maple wings, and cocoa beans. In fact, these woods range from the softest softwood to some excessively hard and heavy which sink in water. The commercial species, however, are generally hard. The hotter it gets, the better the hardwood trees succeed, provided there is rain. The better they grow, the more species there are, so only a small proportion are valued; and as they are "hard," the cost of production is high. So in normal times, these are the woods that are mostly used for high quality work.

A little digression but important to the story.

~ ~ ~ ~ ~ ~ ~ ~ ~ ~ ~ ~

When I returned home from Sweden, the time had at last come to join the family business. It was an important milestone for me but in a way that I did not appreciate for some years to come.

Since the end of the war and while I had been away, the derelict wharf had been transformed. Where the hell holes had been, there was a brand new sawmill and timber sheds, and the business was booming in the post-war flourish to rebuild and recover. The only restraint was Britain's restraint – the country had no money. German and Austrian softwood was being imported under war reparations, but the mainstream supply from Scandinavia or Canada needed hard currency. As importers we had to have a special licence, but more than that, our customers, builders, had to have a licence to use softwood in their houses. There certainly were difficulties.

But, one thing Britain did have was her Empire. Admittedly not many softwoods, the workhorse timbers of the world, occurred in the Empire (except Canada which did not count, as it was not in the Sterling area) but there was a big reservoir of hardwoods. So, just as I came back something special was happening in the timber trade. The supply gap of softwoods for construction was being filled by a mystery package of hardwoods.

When I set eyes on the wharf for the first time as an employee, I saw a surprise. Mixed with the stacks of creamy-yellow neatly sawn softwood, as I had been used to in Sweden, were shades of red, brown, grey, and green. But it was not only the colours that caught the eye. There were very wide boards, "waney" or still with the bark on the edges, and huge logs up to six feet in diameter and 30 feet long. Even that was not the sum total of the surprise. These woods had unfamiliar names such as iroko, sapele, utile, meranti, obeche, and keruing.These are names that are fairly well known in hardwood circles now. But other intriguing woods had even more intriguing names such as panga-panga, sanga-sanga, tanga-tanga, banga-wanga, jongkong, logagbola, coigue, opepe, and owewe.

Every week another strange wood with another strange name came into the yard and was pounced on with Cockney enthusiasm and humour. "Cor, wassis, owewe; it stinks" and it did. Then there were the other aspects of interest – how heavy is it, does it keep straight as it dries, is it very rigid or does it bend conveniently making it easier to use as a lever when stacking and carrying? Then, of course, there was the interest of taking pieces into the mill, sawing and planing them to see what they looked like.

I was fascinated, I was hooked. In no time I was touring the miles of mystery woods cutting small pieces of tepa, coigue, abura, nyatoh, idigbo, danta, and Chilean laurel. These I took home to plane and examine. Larger pieces I experimented with by making fairly basic pieces of furniture.

Soon I had small pieces of quite a few woods – perhaps about 50. They were all sorts of shapes and sizes and all sorts of colours, densities, and textures. I found myself getting quite ecstatic about some of them. My enthusiasm seemed to be infectious – not that other people at work wanted to collect pieces too, but they seemed to want to find more new woods for me. The salesmen in particular brought back pieces of woods that other importers had found. Slowly it dawned on me that I now had a timber collection. It was a fairly scruffy collection but a timber collection. The pieces were of all different sizes with names scribbled on them.

By now I was making contact with hardwood importers, not just softwood importers who were making experiments. I paid visits to the hardwood yards, mostly in the East End of London (and mostly now gone), and they gave me pieces of teak, beech, and many kinds of oak, mahogany, and walnut.

Then somebody told me about a Dr. Desch, a wood technologist (that sounded impressive) who sold samples of timbers through a book-selling company. Post-haste I was in touch with them. They sent me a list which included at least 50 woods which were new to me. In no time they were in my collection. They were not only attractive as being yet more new woods to look at, but there was something new – they were all one size and each had a nice adhesive label which gave the trade name, the botanical name, and the part of the world from which it originated. I realised what a jumble my collection was, made up of such a hotch-potch of pieces. I resolved to put this right. Dr. Desch's samples were 4-1/2 inches long, 3 inches wide, and 1/2 inch thick (which, as it happened, were modelled on the South African Forest Department standard size); I took the step to conform to this, which was to decide the standard size of my large collection.

By now I had about 200 woods in my collection. Most were over the size I had chosen as standard and I could easily bring these down to it. But, quite a few were smaller. So of course I had to go back over all that ground and get better pieces. By now I had become used to the expressions on people's faces when I asked them if I could have a sample of a wood. But to go back again and ask for a larger piece was almost too much.

Even so, it was not long before all of the pieces were one size and I saw great benefits. Each piece could now be compared easily with another, storage was much easier, and it began to look like a businesslike collection. Now I typed very basic labels for each piece, and they could be stored on shelves in our old playroom at our house which conveniently adjoined the workshop. Next I took the final formative stage – I started a catalogue which showed the botanical family to which the tree belonged, the particular species of the sample, the best established trade name, the area of origin, and a three-letter code which could be used to consult a list which showed the full name and address of the supplier of the sample. I suppose I was now really organised. The ground rules have hardly changed since then. Perhaps the only two refinements I have added are that each sample is hammer-stamped on both ends with the accession number, and I have printed adhesive pro-forma labels in strips onto which I can type the required information, cut the label with a guillotine, and stick it to the new sample.

Still the samples came in. By now I had discovered the creme-de-la creme of the hardwood trade – the exotic timber merchants as they were called. At last I had met the real connoisseurs of wood. Not just the people who sold these exquisitely beautiful woods but their customers who lovingly made gorgeous masterpieces of furniture, carvings, turned bowls, and sensual sculptures from them. All shared the drooling admiration for a perfectly beautiful, natural product. I felt privileged, and yet totally at home to be among them. Now the collection was enhanced by such gorgeous woods as Rio rosewood, Brazilian tulipwood, kingwood, boxwood, cocobolo, lignum vitae, and different kinds of ebony, not all of them pure black.

I now had 297 samples, and I was enjoying myself. It was May 1954 and my National Service call-up papers came.

Chapter 6
For Queen and Country

Above the town of Hednesford, in Staffordshire, England, is a large area of unspoilt woodland. Only two sheds are concealed among the trees. It is a scene of sylvan tranquility. It was not in 1954.

It needs a lot of imagination to bring back to life the Royal Air Force station Hednesford, a National Service Basic Training Station, much better known as a Square Bashing Camp. To this place I was brought as one of a busload of about 30 apprehensive young men those 55 years ago. Where there are now birch and pine trees, there was a guardroom which seemed to spawn some of the most aggressive people I have ever met. Further back was a range of administrative buildings and beside these a huge paved parade ground. Ranged around that were many rows of wood huts, and all of this was surrounded by a perimeter fence which enclosed many acres. This was to be our prison for several weeks as we were brow-beaten into subordination. Each of the wood buildings housed about 20 raw recruits, and in a separate part of the hut (or billet as it was called) lived our particular tyrant, the Corporal Drill Instructor or D.I. Just what kind of environment could have bred such sadistic people is difficult to imagine. Their power was absolute. It began early in the morning when the recruits went off to their ablutions, polished the floor, set up their blankets and sheets in a ridiculous cubed pattern on their beds, and went off to breakfast. It only finished with lights out at about 10:00 pm by which time polishing (or bulling) of boots and all the irritating brasswork of uniform had to be complete. Somewhere in between, these tyrants yelled semi-intelligible screams and exhortations as we marched up and down, left and right, slow and quick, as some of the many groups of victims on the parade ground. Those who presented the slightest challenge or committed the smallest misdemeanours were sentenced to fatigues such as washing out lavatories or ordered to have their hair cut, no matter how short.

For the first half of the six-week time, we were confined to camp. The only possible escape was in the evening to run the gauntlet past the guardroom to a fish and chip van parked outside. I plucked up courage one evening to do this, hoping and praying that the machine-gun "Airman!" would not be shouted at me. I did not want to get any greasy marks on my uniform as I walked unobtrusively back with my fish and chips, so I held them well in front of me.

"Airman!" I walked briskly up to the gloating corporal.

"Yes, Corporal."

"Were you a waiter in civvy street?"

"No, Corporal."

"Do you realise you look a stupid idiot – you are a stupid idiot."

"Yes, Corporal."

"And get a haircut."

"Yes, Corporal."

It got better as time went by. We were allowed out to nearby Cannock for some evenings at the cinema or the pub and a wary kind of teamwork grew up between us and our tyrannical corporal. One day we were hustled into a large room where on one wall there was a list of all of the trades or occupations available to us. None appealed much to me. There was nothing exciting like a wood aircraft propellor maker. I did not want to be a fitter or operator or clerk or, the last resort, Admin Orderly, which was an out-of-character euphemism for dogsbody. There was, however, something called a Fighter Plotter. I had little idea what that was but it sounded intriguing. I found later that most people doing this work had chosen it for much the same reason.

"You stand by a large map on a table with a magnetic fishing rod in your hand, and move arrows across the table to represent aircraft movements," they said, "while somebody sitting in front of a radar screen tells you on the earphones what to do. You both work for two hours and need one hour's rest because it is not good for the radar operator to stare at a screen for hours." This sounded decidedly promising. I knew for sure I had made the right choice when they said the training course was to be at a place with the lovely name of Middle Wallop, in Hampshire.

Freeing myself at last from Hednesford I found myself safely at Middle Wallop and discovered that the rest of the Air Force was not at all like square bashing camps. The atmosphere was congenial and civilised, and here I had the pleasure of meeting Robin Martin who was one of the most interesting characters I have ever met. He was the son of a successful lawyer and later became a very well established lawyer himself but, sadly, died several years ago.

At Middle Wallop, he had decided to exploit National Service, making use of every pleasant opportunity that might present itself while artfully dodging every unpleasant duty. Whatever parade was called, whatever fire picquets or guard duties there were, there was always a reason why Robin did not have to do them. He was so much more intelligent and resourceful than his masters and his face was one of such pure innocence that his plots never failed. I felt that Robin was the best possible friend to have – besides I found him highly amusing.

The course was easy, and most, including Robin and me, passed the exam at the end. After all, the scope for placing arrows with a fishing rod was somewhat limited. In spite of that, two people managed to fail. Both had opted to go abroad when they were called up. So had Robin, ever keen to work the system. For some reason I had not. Now they called for two volunteers among those who had passed. On impulse I put my hand up. So did two others. My name was the second pulled out of a hat.

Now I was going "abroad." I had no real idea of what "abroad" was and nobody seemed to know exactly where it was. But they told us that it began with two weeks embarkation leave, and when we got there our pay would be doubled to £2 a week. Abroad sounded good.

Back from leave I was told to report to another RAF station and Robin and I learned that for us abroad meant the Far East. In a busy week we were given another set of uniforms. This time it was a "tropical kit." We were vaccinated for just about every dread disease and were pretty sore about it. We saw films about how to land and live in the jungle, the desert, and even the Arctic, and they gave us lectures about how to avoid all sorts of sexually transmitted diseases.

At last we were ready, and slowly the secrets came out. We were to go to the port of Southampton and board the troopship Asturias. There we eventually learned that, for Robin and me at least, abroad meant Singapore. This has changed for me now, but geography had not been my strong subject at school so Singapore did not mean much to me.

Asturias (God bless her recycled steel) was far too good to be a troopship. She was quite big (27,000 tons) and was a sleek, white-painted ship which could move along quite niftily. I don't know her history. I didn't even know then that Asturias was a district of northern Spain. I just climbed aboard this vessel and found myself in berth no. 140D on E Deck (I still have the boarding card). There were four bunk beds in this cabin which made it luxurious as we were told that, unlike the pampered Air Force, Army people slept in hammocks in large open areas.

The late afternoon saw us slipping gently out of Southampton, virtually unnoticed, into the gathering gloom. It was mid-December and we would have this ship as our home for three weeks.

In the night, crossing the Bay of Biscay, I again found my stomach protesting violently at the rolling and pitching. This was not made easier by the ever-creaking and swinging doors, and the gurgling bowls of the lavatories. But within a few days even the worst sailor got his sea legs and we were never troubled again.

All of the next day we passed by the spectacular green hills and mountains of Portugal and soon we sailed past Lisbon, which I remember as a glorious splash of white buildings against the green background. As dawn broke the next day, Gibraltar came into view. As anyone who has sailed past Gibraltar that way will know, it is a tremendous sight. The west side is an almost geometric 45-degree slope covered with woodland and nestling at its base is the town. The east side is almost perpendicular rugged crags. All too soon it slipped away behind us, but the rest of the day we cruised past the coast of Spain. The scenery was spectacular. As it was mid-winter the mountains were covered with snow. Many years later I was to explore these mountains and climb the heights of the Sierra Bermeja and Sierra Nevada. Down at sea level were little fishing villages such as Torremolinos, Marbella, and Fuengirola.

Gradually the land receded and for the first time we felt the sensation of being completely free in our own island. Asturias seemed to be racing through the seas as though she would never stop, and we got used to the uncanny feeling of accelerated walking as we strode forward on the scrubbed decks. Everywhere was gleaming white paint and that special smell which ships always seem to have. Alongside us hundreds of dolphins cavorted – this was the good life for everyone.

There are several things that were left unsaid about National Service. The most important one was that there were simply too many people to do everything. Jobs had to be invented or elaborate rosters created which gave the impression to people, who did not actually have to be deluded, that everywhere was a hive of activity. So, casting loose a thousand fighting men or so, for a leisurely three-week cruise, was really a convenient and cost-effective way of transporting to the outposts of Empire. Nominal duties took place before breakfast, fire picquet patrols, more films about horrible diseases, film shows, and some impromptu turns on the stage. But most of the time was spent on Tombola, marathon card games, or "guess the ship's mileage for today."

We passed between Sicily and Malta and before long we dropped anchor at Port Said, ready to go through the Suez Canal. For most of us impressionable youngsters constrained to Britain by the war, this was our first experience of the orient. Unfortunately, we were not allowed ashore because we were in uniform and this was considered unwise in the delicate political climate. But even standing at the ship's rail, the sights, sounds, and smells of Egypt wafted over to us. Over to us, too, came a small army of bum boats which were small floating bazaars. In no time, ropes were aboard and goods were being exchanged for cash by a very effective pulley system. Soon these traders were aboard. In one room the Gully-Gully man had set himself up. He was the most accomplished conjurer I have ever seen. In the space of a room he was pulling rabbits out of hats and practising every kind of deception, punctuated by "Gully, Gully, Gully" until we stood spellbound.

As dusk descended most of these pirates were hustled off again as we joined the queue to head south through the Suez Canal. A few remained, however, to frighten us in dark corners saying "Do you want to buy picture of my sister, or do you want to buy flick knife?"

When we woke up the next morning, it was to a surreal sight. On both sides of us there was sand as far as we could see – and there was a ship plodding across it. Actually this was not a mirage. At that point in the journey the canal is up to motorway or interstate standard as one might say, with widely separated roadways in each direction so that it looks as though the other ship is coasting across the sand.

Past the Great Bitter Lake and onwards to the Red Sea, it began to warm up noticeably. Before long the message went out that on the next day "tropical uniform will be worn." On that day there should have been a knobbly knees contest instead of guess the mileage. We Air Force people were now wearing sand-coloured shorts and tunics which were to become our habit for at least the next 18 months.

When we reached Aden, at last we were allowed ashore – the first footing into the mysterious orient for most of us. The scene was stunning. Familiar as we were to a northern well-watered land, instead we saw totally barren red hills, gleaming white buildings, Arab cloaks and hats, groups of recumbent camels, and a mild air of menace. But we landed at its little port of Tawahi and took comfortable American taxi cabs to the centre, Crater, where we began to learn the refined art of oriental haggling.

Soon we were on our way again down through the Gulf of Aden and out into the Indian Ocean. Now it was really hot and the sea was treacly smooth. An air of lassitude and goodwill seemed to waft over us – just right in fact for Christmas Day. I still have the menu of that memorable Christmas lunch aboard Asturias "at sea." Served to the troops by the officers in the best traditions, the food departed not one iota from its tradition of tomato soup, roast Norfolk turkey with stuffing, and Christmas pudding and rum sauce. Father Christmas appeared sweltering in full regalia; there was some in-house entertainment and then we sauntered out to swelter, for ourselves, on the sun-blessed decks. Somehow from then on all of the service things linked to home, like square-bashing camps, seemed to recede with the backwash of the ship, and the tempo and flavour of the east and of tropic heat and humidity seemed to be in the air ahead of us. We stopped at Colombo, capital of what was then Ceylon, a bountiful verdant and welcoming city, and then set course further east for the last stretch.

I shall never forget our arrival at Singapore. All of the best things on that voyage seemed to happen in the early morning. As we awoke that morning, we saw the most beautiful tropical scene one could imagine. On our left was the mainland of Malaya, every inch covered with rampant green. On our right were hundreds of low islands, each with its clean, creamy white beach, and each clad with coconut palms, some with just one. As if to provide full cruise facilities, Asturias seemed to be coasting along at well below her usual speed. The islands drifted past us, and so did small fishing boats manned by happy men who waved and smiled at us. Singapore looked very promising. Eventually we came to a halt at the entrance, we were told, of Keppel Harbour, where we were to wait some time for a berth. There was a carefully controlled air of excitement as we thronged the decks, and Asturias finally and slowly manoeuvred into the dock. The first thing that happened when we finally berthed was that a small number of officials came aboard, followed by, to a rousing cheer, a much larger number of Chinese women in black pyjamas. If this was to be Singapore, we said, it looked even more promising. But they were cleaning ladies, come aboard to attend to the ship's linen.

At last came the time for us to disembark. Picking up two kitbags (one ordinary, one tropical) we stood, a rather dishevelled lot, in ill-formed ranks on the quayside. It was now mid-morning and the sun was beating down on us. It was then that we realised what true tropical or even equatorial heat (Singapore is only 90 miles north of the equator) is really like.

In Singapore, we were told, the temperature has never dropped below about 75°F and never goes much above 95°. In fact it is in the mid-80°s most of the daytime I think. As I write I have just looked in our daily paper and seen that yesterday's temperature was 89° and it was thundery. It is not, therefore, the temperature itself that strikes you, but the combination of that and the incredible humidity. On the voyage I had brought with me a raincoat, as I was told it rained a lot in Singapore, but one of the old hands told me I would never wear that and it would be rotten in three months – he was right; it was in shreds in a cupboard in that time. Any self-respecting serviceman carried a waxed paper Chinese umbrella to work.

As we stood on that quay wondering what would happen next, I think we also wondered whether we could actually cope with that climate. The sweat poured off of us as we simply stood there with no exertion, and the atmosphere was drenched in humidity like some all-pervasive inescapable Turkish bath.

It was not long before we were herded onto a bus for the Royal Air Force station at Changi. Only then did we know that this station was to be our home. This journey inevitably took us through the City of Singapore. It was one of the most spellbinding experiences of my life. I had never seen anything like it.

There was an extraordinary array of buildings – colonial majestic, ultra modern, Indian, and Chinese with covered alleyways for shade lining the streets, hovels, the Cathedral looking like some amazingly misplaced folly, and the totally in-keeping Raffles Hotel; signs in English, Chinese, Indian, and Malay, men and women of every hue dressed in every dress, aromas of every kind from the searing curry to the luscious and sensuous Frangipani tree. Especially I remember the impact of the trees. I did not know their names then but there were trees extravagantly covered in hordes of bright red flowers, trees with pods two feet long, palm trees whose trunks grew fatter near their tops than at their bases, trees like palms but with leaves spread out like a fan, trees with huge stinking fruits, and trees whose leaves curled up in the rain, to collect water. I was spellbound.

Out of the city we moved on north-eastwards past ramshackle villages and narrowly missing equally ramshackle buses and trucks driven by maniac Sikhs and Chinese, until we neared Changi.

First we saw Changi Jail, infamous as the British Prisoner of War camp run by the Japanese in the war, and then at last we pulled up in front of our new home, Block 140. Changi is set right at the north-east corner of Singapore Island, with the sea on three sides. The view from Block 140 was idyllic. Before us was the runway where every keen airman could see what really mattered, happening each day. Beyond that were the beaches, the sea, and those palm-clad islands. Around the block were well-mown lawns with those gorgeous trees – frangipanis, flame of the forest (the ones covered in red flowers and with huge bean pods), traveller's palms (not really palms at all but having palm leaves spread out like a fan), mangoes, coconut palms 70 feet tall, and giant forest trees with trunks like marble columns three times as tall.

The block itself was magnificent. It was built in the style of an Indo-Malayan palace on three floors. The windows were not glazed but were open spaces between colonnades ranged around the entire building. Inside these were pleasant balconies, and inside these were louvre shutters (which were never closed) to the sleeping areas. Each floor accommodated 50 airmen. It may sound a bit penal now but to us it felt like paradise. The block was undoubtedly the best sited of a small number. Next to the block was the one with the cross on it. It had been the hospital for prisoners of war of the Japanese, when the drains were said to have "run with blood."

The scene was, mercifully, happy now, and as we walked to our beds on the middle floor in the late afternoon, the atmosphere was very welcoming. Above our heads the fans, six feet in diameter, wafted cooling air down to us. Good sport we later learned was to throw berets into the fans and lay bets on which direction they would be projected. On the walls, chitchats (lizard-like animals) clung with their padded feet waiting to catch insects while giant mobile beetles hovered.

Despite the fans, it was still very hot and several people were lying on their beds wearing little and in some cases nothing. It was, therefore, a great surprise to me when I saw a Chinese woman sauntering totally unperturbed through the large room muttering "so so." Actually it was Sew Sew – she was the sewing lady who made a living plying the camp as an itinerant human sewing kit. If it was not Sew Sew it was Mary, a diminutive Chinese of incredible strength who carried two vast baskets of fruit, so heavy that no one else seemed to be able to lift them. Or it was Ice Kleem, the Magnolia Ice Cream man who came with his freezer box, and the Dhobi man who called weekly for the laundry. None of them spoke a word of English except their wares and dollars and cents. If teased about money, Sew Sew in particular would round on her oppressor with a vehement unintelligible tirade. She was always paid in full and tipped generously. But the real luxury was that even the most humble "sprog" had a share in a bearer who made his bed, cleaned his shoes, swept the floor, cleaned the washing areas, and went on errands to the camp shop. For this he was paid a small amount a week by each of us, which meant that altogether he was actually quite well paid. Our bearer was Parla and he was a gem. He was a most delightful Indian who pulled out with the Raj in 1947 when India gained her independence and was thoroughly used to the ways of British servicemen. His ever-cheerful manner was a delight to us all.

Chapter 7
Outpost of Empire

Fate had decreed that Robin Martin and I should share all of these experiences. Since we had the same adventurous but carefully selective outlook on life, we also became the best of friends.

Our first morning at Changi, after breakfast at the mess which served traditional English and daring oriental food (and where we in Indian style were to have tiffin – never lunch), we ventured to find our place of work. It was the radar station with the imposing name of the Southern Sector Operations Centre. Changi was said to be the largest Royal Air Force station in the world. The radar station was by a beach on the other side of the runway, totally isolated from the rest of the station. To reach it, we travelled on ordinary Air Force lorries which were called gharries (another Indian colonial survival). Equipped with primitive radar aerials, its function in life was, I suppose, to look out for possible but very unlikely intrusions by Indonesian aircraft. Inside it looked like another Middle Wallop and in no time we were at home. In the main room was the huge map on a table and in separate rooms the radar operators spoke things through earphones to us fighter plotters as we marked aircraft on the map with little arrows. Overseeing everything was the Fighter Controller who no doubt longed for the day when he could scramble a squadron of fighters (or even just one) to intercept an intruder.

Whether it was a medical fact or whether it was a convenient way of coping with excess numbers, two hours were decreed to be the longest time that a radar operator could look at his screen. Then he needed an hour's rest. To make it fair, the same applied to fighter plotters, and after the two-hour spell, we would be released into the open air. This was the signal to adjourn to the tea-swindle which was a pensioned-off gharry turned into a sort of café, run by a fortunate radar operator or fighter plotter. Here were planned the enthusiastic outings for days off, and the service-weary ticked off months, days, hours, minutes, and seconds to repatriation and demobilization.

On days off there was plenty to do. Changi Village, comfortably contained within the station, existed almost entirely as a bazaar for service people. There were a number of shacks or sheds with such names as Glamour Silk Store (Always at your service, Ladies Dressmakers and Gents Outfitters NB we despatch parcel to UK), Curly's Evergreen Store (the only shop in the village where you can be fully satisfied), Cheap Jack (your favourite shop), Cash-n-Carry (Best selection of China-wear, nylons, musical boxes, and watches, etc., etc.). Here you could polish your bargaining prowess as you loyally bought things for the folks back home. (How loyal we were!)

Changi also had a nice swimming beach where you could see strange crabs and pick up wonderful coral. Out toward the mainland was Ubin Island to which you could swim or take a small boat. You could walk around that island in an afternoon and see growing coconuts, pineapples, papayas, mangoes, jambu lauts, and the evil-smelling but nice-tasting durians. All in an afternoon you could also see small crocodiles, rubber trees being tapped, and walk through a typical seaside village or kampong before heading back to the ferry beach which was overrun by feral guinea fowl.

I didn't know it, but apparently Changi was considered to be the finest RAF station in the world with every facility you could wish for – football and cricket grounds, a golf course, a swimming pool, a good library, the Malcolm Social Club, and grounds with ornamental trees like a huge park. Today, this is all replaced by Singapore's Changi International Airport, one of the finest in the world. It was inevitable that this has been at the expense of the old RAF station, which has virtually disappeared now.

But in those bygone days most of our outings were to Singapore City, that magnetic centre of attraction. There was hardly anything you could not do in Singapore City. Passing through the Cathedral grounds you could pay a few cents to the Indian snake charmer who in return would open his box, play his pipe, and lo and behold the snake rose gently and posed elegantly to the entrancing music.

At the south end of the city was the Tiger Balm Garden. Tiger Balm, as every good Singaporean knows, is a wonderful panacea in the form of a spicy ointment in a small round red container which everyone should have in his house. It cures, the label says, coughs, colds, headaches, rheumatism, gout, sciatica, lumbago, sore throat, toothache, asthma, scorpion and other insect bites and stings, cuts, bruises, cramps, and all chest complaints. To prevent infectious diseases, it is advisable to keep a tin of this balm when travelling and at home. So highly esteemed was it that the Haw

Par family had made a fortune out of it and decided to build the bizarre Tiger Balm Gardens. After walking under an entry arch in the form of a Chinese pagoda, one passed an array of weird figures of tigers chasing snakes, orangutans, a huge figure of Buddha attended by a sweet Chinese lady, tortoises standing upright and riding on emus, and a rabbit shooting a cat with a revolver.

With all of the trees around me, wood collecting could not remain in the background, and before long I found my way to Singapore Botanic Gardens. Now I could put names to the strange trees I had seen in the streets. Here for the first time I saw a balsa tree and a teak tree. In the centre of the garden was an administrative building and seated at a desk was a studious and official but friendly looking man. Soon we were talking animatedly and the subject of wood came up."Yes," he said, "I will be very happy to get someone to put aside for you some logs of trees we are cutting back or down for various reasons. Come back in a week and we'll see what we can have ready for you." In exactly a week I was back and there waiting for me were about 20 logs, about a foot long and up to a foot in diameter. Each was a species new to me. I was thrilled and extremely grateful. I was also keen not to show any anxiety about how on earth to get them back to Changi and what I could do then. Ever-helpful, the staff got out a large felling axe and split out a wedge-shaped piece of each for me. These I managed to haul home to Changi and there at the end of Block 140 I rigged up a small vice and sawed them down to a little over my standard size. I still have them. I had to buy the tools and this was another of those times that I had to shrug off the incredulous "what are you doing?" I also had to pay Parla a bonus for helping me clear up the mess.

"If you come back again next week we could probably find some more spare samples of the timbers from outside Singapore and Malaya which we have somewhere" said the friendly man at the Botanic Gardens. So of course I was back again ... and again. There were several from British North Borneo which is now Sabah, a few from India, and most interesting of all a small number of samples of timbers of Formosa labelled in Japanese. Formosa, now Taiwan, was in Japanese control during the war and the samples had, strangely, been brought down to Singapore by the Japanese. My cup overflowed.

"I can give you some addresses of Forest Departments of governments of other countries – "why don't you write to them?" said my good friend and mentor. "Start with the Malayan Forest Research Institute at Kepong, and then try India which has a legendary reputation for forestry, and how about Australia?"

So, first I tried Kepong and received a most helpful reply. "You can have about 50 free of charge in our standard size of 4 by 3 by 1/2 inch," they said. This was most welcome but also a problem as my standard size was 4-1/2 inches long. "Never mind," they said, "we will cut pieces specially for you for a small charge." This they did. And I was deeply grateful as it must have involved a lot of time and cost.

Meanwhile, I had made contact with the famed Forest Department at Dehra Dun in India. More splendid samples came from them including rosewood, teak, ebony, and other beautiful woods well known in India but not elsewhere. Every sample had the botanical name hand-painted in its side. Soon after that, replies and woods came from Tasmania, Melbourne, Adelaide, Sydney, and Perth.

Partly because I felt these organisations would have a soft spot for a serviceman overseas, and partly because I wanted to see these samples, I asked for them to be sent to me in Singapore. Soon, as well as the samples coming to me, there were samples on their way home to England. The collecting was going well.

My new good friend, Mr. Burkill at the Botanic Gardens, now suggested I should contact the nearest Forest Officer on the mainland of Malaya. I had developed a keen interest to see some of the trees which produced the woods now in my collection. Once again I had a marvellous reception. So on one day off I decided not to go to Singapore City or Changi Beach; instead a smiling Malay named Bujang picked me up in a forestry jeep at Changi early in the day and took me across the causeway into Malaya for the first time. We were to visit the Panti Forest Reserve in the State of Johore. Soon we reached the forest office. There I was issued with a carbine and some rounds, and we were joined by several armed Malay soldiers – this was risky terrorist country. We set off on a tricky unmade road up to the hills. All around us was the archetypal tropical rainforest – palms, huge tree-ferns, lianas, and forest giants towering unbranched above us. This reserve was being exploited under close Forest Department supervision on a selective felling basis for timber. This was yet another new experience for me as I felt the thrill and fear of seeing mighty trees tumble. This was before the days of chainsaws and all of the work was done laboriously by Chinese coolies with huge two-handed saws. Many of the woods are quite hard so the effort in the humid heat must have been tremendous. As for me, I just stood and watched, and I even got some new samples from the visit. Our only alarm and excursion of the day was when a loud bang was heard in the valley below us. The Malay soldiers cocked their rifles,

A visit to the Panti Forest Reserve in the southern Malayan rain forest, 1955. The area was infested with communist terrorists so there was a guard of several Malayan police. I am armed with a carbine for the day.

looked anxious, and disappeared below. Soon they were back. One of the tires of a logging truck had burst in the baking heat.

After the visit we met the manager of a sawmill and went to his house for a marvellous curry while he showed me the loaded revolver he always kept with him since his brother was shot by communist terrorists. At the end of this marvellous day, Bujang took me back to Changi where I returned feeling like a king and was on a "high" for days.

Chapter 8
Up Country

Meanwhile back at the camp, fate intervened in rather a nice way. While "other ranks" were ferried to and from the radar site in the gharries, the officers had the privilege of traveling by Land Rover. But, of course, they did not drive themselves; they were chauffeured, and not by an MT (motor transport) driver, but by a fighter plotter or radar operator. Miraculously, the drivers of both vehicles were going home, so there were two vacancies to replace them. Would you believe that Robin and I were the only two people with driving licences? So we got the jobs. The onerous work consisted of collecting the vehicles in the morning, taking the officers or sergeants from their mess, to work, transporting them to and fro for lunch, and returning them in the evening. This made for plenty of leisure time. The best time was in the evening when, all work done, we would set off in opposite directions and drive around the huge station until we found each other.

Of course all good things must come to an end and this life was far too good. After six months, one day we were told that a new radar station was being opened in Kuala Lumpur and some radar operators and fighter plotters were needed to start it up. I was one of the "lucky" ones chosen. Robin was not. I felt very fed up at first because life was so good and Singapore was so exciting, but "tomorrow to fresh fields and pastures new."

There would be many differences between Singapore and Malaya, and terrorists were one of them. In Singapore, National Service was almost a vacation; in Malaya, however, it was all about fighting a war. During the World War, up to 10,000 terrorists, mostly Chinese, had been seeking to overthrow the Japanese occupation. After the War, they changed their campaign to seeking to impose a Chinese communist-style regime while Britain was preparing to give Malaya its democratic freedom. Things got pretty nasty with the murder of rubber planters and tappers, as well as rice and pineapple growers. Villagers were being shot and intimidated into submission. By 1954, the Malayan and British forces were beginning to redress the balance, but there was still a long way to go and it was a dangerous place.

In 1954, the Malayan railway system ran on a narrow gauge. The steam trains with wonderful saloon carriages plodded along at 30 miles an hour preceded by a sort of tank on wheels to detonate bombs or mines on the line. They were sitting ducks for the terrorists.

One evening in June I boarded the train at Singapore Station, crossed the causeway, and began the all-night plod of 250 miles to Kuala Lumpur. Like all other service people, I had to take my turn on guard duty with a loaded rifle standing at the open end of one of the coaches. Nothing terrorising happened, however, and my only memory is of the sickly night smell of latex oozing out of rubber trees as the train sedately wound its way through the country to Rengam, Segamat, Gemas, Tampin, Seremban, and, in the morning, Kuala Lumpur.

Air Control Centre Malaya, the new radar station, now had its radar operators and fighter plotters but there was one thing missing – the radar. That was not to come for six months and was then so hopelessly obsolete and worn out that is was only to be used on special occasions such as an official visit.

But it was not a farce. Even if it had its huge map on a table which was never used and fighter plotters' magnetic fishing rods which never placed an arrow, it did have radio – three channels: Lumpur Control on which we listened, Locust on which we broadcast, and the international distress frequency on which we both listened and broadcast, but fortunately not too often.

Our job was to monitor all of the service flights over Malaya every day, record take-offs and landings, and act in a hurry if anything went wrong. There was a surprising array of aerial response to the terrorist menace. The Royal Air Force had contributed twin-engine medium-sized Valetta aircraft which had their doors removed so that supply-drops of food and arms could be made to army hill forts and jungle camps, and the good old Dakotas which had big generators fitted inside their bodies and loudspeakers dangled under their wings to broadcast seductive surrender messages. Also there were Pioneers which could land and take off on a postage stamp it seemed (actually only 75 yards). There were Sikorsky helicopters and modern Canberras used in photo reconnaissance. The Navy had some helicopters too and the Army some Austers, while the Australian Air Force had some Lincoln bombers, survivors of World War II, based in Singapore. Perhaps the most exotic were the RAF's Sunderland Flying Boats also based in Singapore, which stayed airborne for an interminable time while their crew lobbed 20-pound bombs out of the fuselage.

Somebody somewhere had the job of coordinating this melange. Our job was to follow the day's events until each mission was safely accomplished. That was the theory. The practice was rather different. The army felt no need to keep us informed of its movements as its Auster aircraft were so small. Singapore-based movement details somehow rarely reached us. Even some of the Kuala Lumpur movements eluded us as, once the helicopters were weaving their way through the mountains, they could not maintain radio contact and some honoured the code only in the breach. All of this was fine until anything went wrong. There were mercifully not too many bad moments, but when there were, it was a very taxing time.

Particularly with no radar to operate, or fighters to plot, Air Control Centre Malaya was even more grossly overstaffed than its opposite number in Singapore. The work of a normal day began with a listing of all of the flights for the day on a display board called the tote. As the flights took off, the monitoring, communication, and recording continued. By the evening, all of the flights had been concluded and all that remained for us was to take the details of the next day's flights, notify Singapore and Kuala Lumpur airports of danger areas, and listen on the distress frequency.

A most satisfactory shift system had evolved which catered admirably to the big excess numbers and yet gave the impression of full commitment. (I suppose with hindsight that the large numbers were not as bad as they seemed, since they would be needed to cope with an emergency). On the first day-shift, one worked from 2 pm to 8 pm, on the second from 8 am to 2 pm, and from 8 pm to the following 8 am, when they had the day off. In fact only half the shift worked at night but they still got the day off every third day. No officer ever attended a night shift but occasionally rang in to see if everything was all right. Even for the worthy-two working the night shift, life was easy because by ten o'clock all of the next day's movements were known and listed. A phone call to the civilian airports established that all of the flights for the day were complete and all you had to do then was ring the station telephone exchange and book an early call for six o'clock. Then, you could go to sleep, even if somewhat uncomfortably, on the table-map.

I excelled myself one evening. My colleague for the night had gone down to the mess for a late meal. I was on my own, it was quiet, and the aircrafts had all landed. On evenings like this, as a 19 year old I would write letters home and send drawings I made of the strange army or air force of nocturnal insect visitors which would be one's only company. I had a favourite large beetle named George, and various other fliers would appear including huge and ferocious-looking locusts. But tonight George and his friends were elsewhere, nothing was happening, and I was bored. Someone had told me that by pressing a few buttons Radio Malaya could be played in the room through one of the loudspeakers. Soon I had some jolly music for company. Soon also I had a phone call from the airman manning the VHF/DF homing station.

> "Can you hear music on the distress frequency?"
>
> Like a fool I said, "It's me." I rushed to correct my error and turn it off.
>
> "I shall have to report this in my log" he said.
>
> "For God's sake don't do that" I pleaded.
>
> "Well I could say I stood down for five minutes for a pee" he replied.

For the next few days I lived in mortal fear of the most horrendous court martial but nothing ever happened and that man on the homing station and I have kept our secret until now.

The fun way to spend the many days off was to go on flights over the jungle. We were on such good terms with the Operations Room at Kuala Lumpur that they would readily get us on all kinds of flights. Nobody could be a passenger, always an observer. The more adventurous of us regularly

A Royal Air Force Valetta puts on a display over a Malayan river after making a supply drop to troops in the jungle, 1955.

flew on all kinds of flights. The Valettas were fun and the missions quite hair-raising as these medium-sized, only moderately agile planes manoeuvred around mountains to drop supplies. Once I remember we strayed over Thailand for a visit but I am sure no one cared or even knew. The little Pioneer and helicopter flights were fun too, as they went into the jungle camps and you met aborigines as well as Malays and British. Once I took control of a Dakota for a few minutes – a daring liberty by the pilot.

On one occasion as I put in an appearance for a Valetta flight, the atmosphere was cool. There was a very upper-crust army officer who had come along for an observer flight.

> "Of course you chaps don't have any idea what it's like down there, up to your waist in muddy water, leeches clinging to you, terrorists behind every other tree!"

We took off. The commentary continued.

> "If you were down there at that dropping zone you would really know what it's like to be fighting a war out there."

Not a word from the pilot or navigator – only a look passed between them. All of the supply drops were completed efficiently and normally. But then we headed out to sea. First we shot past the only multi-storey hotel in the lovely old town of Malacca so close that we could see into the bedrooms. Next we dived toward a fishing boat so that the occupants were so scared that they jumped out. It is difficult to believe the next thing but we headed inland and actually flew under the cables between two tall pylons. By now I was really scared as I thought things had gotten out of hand. Next we were flying in a narrow valley between two mountains. I knew that most of the disasters we encountered at the Control Centre were related to the aircraft being unable to climb out of these valleys when they were suddenly faced with a mountain ahead. I sat behind the crew and looked in front and saw just such a mountain ahead. "This is it" I said to myself "there's no getting out of this, the pilot has risked too much." Miraculously at the last minute he tugged at the controls, we were forced to the floor, and the plane safely cleared the mountain top. He knew exactly what he was doing.

When we landed there was a very green looking army officer. Still not a word was said. Just another supply drop. But as for me, it was one of the great formative events of my life. What is the use of worrying about something you cannot control?

Chapter 9
Upside Down – and Back Again

Other days off were spent exploring Kuala Lumpur. After pulsating Singapore it was rather different, but its splendid railway station in superbly ornate Indian style was most impressive and the government buildings in front of the Padang (now Merdeka Square) were also very fine.

It was not long before my thoughts again turned to wood. I phoned the Forest Research Institute, and on one of my free days I went out to Kepong. At this beautiful white building set in parkland with palms in front and a backdrop of forest I again experienced really warm hospitality. First I was given a thorough tour inside and out. I saw strange new machines for testing the strength of timbers and a museum containing some horrendous examples of destruction of wood by termites, carpenter bees, and savage rots. But the museum also contained superb panels of Malaya's beautiful woods. Of course, it was not too long before we gravitated toward the wood pile and, in addition to all of the samples they had already sent to me in Singapore, together we managed to add many more.

It was Kepong also which put me in touch with the South African Forestry Department, undoubtedly the most efficient and consistent suppliers of wood samples anywhere at that time. Each sample was painstakingly and accurately prepared and named, and the list contained an ever-growing number of native and exotic timbers which were to be added steadily to my collection over the years. On one joyous occasion some years later there arrived a parcel of over 100 different new samples – one of the biggest hauls ever.

But back in 1955 and back in Malaya, the Forest Department was absolutely marvellous. Perhaps my enthusiasm helped, but I can only think of kindness as any other reason. We paid many visits to mountain forests, lowland forests, swamp forests, mangrove forests, and Chinese sawmills. The Selangor State Forest Officer invited me to his home where among other things he and his wife introduced me to more Malayan fruits such as rambutans (like lychees) and the gorgeous mangosteens which were much the same size but had exquisite pulpy flesh – definitely a connoisseur's fruit.

I made many good friends back at the Air Station as well, and I wish I had not lost contact with them now. One in particular, Gus, had much of the same roving spirit as mine. For most people leave was spent at Singapore or Penang Island and some went to Bangkok or Hong Kong, a few to Japan. But Gus had greater ideas. He had met an English officer in the Malayan Police who, as it happened, was just about to retire with his family to Australia. Only a month or two after that had happened, Gus received a letter saying he would be most welcome to spend his leave in Perth. All he had to do was to find the means of transport. This he did easily by "cadging a lift" on the RAF air bus that used to fly monthly from Britain to Adelaide and return. When he came back, he was full of it and I was very envious, especially as I knew that Western Australia had some famous trees called jarrahs and karris.

The Eaves' responded very favourably to my letter. Yes, they would be delighted to make me feel at home and when was I coming? Sadly, it all looked as though it was going to founder when after two months I could not cadge a lift on the airbus. Fortunately my mother, God bless her, sent me the balance of the airfare beyond what I had saved, which enabled me to fly out on Qantas.

With the delays, time was running tight for the month's trip. I was due to return to England at the end of May and it was only at the beginning of April when I finally set off for Australia. I started at Kuala Lumpur, flying down on a Malayan Airways Dakota to Singapore. The airline ticket included, believe it or not, an overnight stop at the Raffles Hotel. It was a place of much romance and a little less grandeur than now. The fans turned gently in the bedrooms and the chitchat lizards climbed up the wall. The dinner menu cost just 6 Malayan dollars and 50 cents (about 75 pence or 40 US cents) and was nothing more exciting than grapefruit cocktail, minestrone, pancake, roast pork, cherry tartlets, cheese, and Borden's instant coffee. But each table was looked after by a single waiter and the service was splendid.

Up, up, and away early the next morning, the Qantas Super Constellation, one of the most elegant of airliners of its time, took its charges to Djakarta which was then a very unfriendly place. Indonesia was in a state of confrontation with Singapore and Malaya. We were allowed out of the plane for refueling, but we were told that the police would put a bullet through our cameras if we tried to take photographs. I actually took a couple of defiant photographs, and I kept that camera for many years.

Another eight hours brought us all to Perth. What a splendid city it was. Set picturesquely around the mouth of the Swan River (perhaps it should be called Swan Lake), it spread itself gracefully, with King's Park's thousand acres as a central feature. Cinderella capital of a Cinderella state it called itself then. This was an inferiority complex it never deserved and thankfully now has thrown off forever.

The Eaves were marvellous and their welcome magnanimous. Their daughter Peggy and her friends had jobs, so during the day I explored Perth on my own. In the evenings, we had barbies and lashings of thirst-quenching Swan lager.

I had been in touch with the Forest Department before my arrival and with the aid of the Tourist Department, two of my four weeks had been set aside to head south to Western Australia's famous jarrah and karri trees. After visiting a sawmill in Perth and collecting some samples, including a piece of the Coolabah Tree of the song (it is a nice, incredibly hard, nearly black wood speckled with small chalk deposits) and sandalwood, I took the leisurely train south and then the bus, until I arrived at Pemberton, near the south coast.

The Pemberton of 1956 was an engaging little village based around the karri forests. A row of rather basic shops ran down one side of the road, and at one end was the splendid Pemberton Hotel which would have been thoroughly at home in the wild west. It had a magnificent wood-colonnaded balcony all along its front, and to stay there was to feel one was perhaps at the remotest corner of Patagonia.

The karri is one of the finest of the great family of Eucalyptus. In every way it excels. It is among the tallest (up to 270 feet), its trunk is one of the straightest and most

The wild west style Pemberton Hotel in the forestry village in southern Western Australia, 1956. Ties and jackets were de rigueur for all meals.

free of branches, and its bark is completely smooth. To walk through one of its forests is to walk through a forest of marble columns. Its wood, which is a rich red, is one of the finest and strongest and was available then in huge sizes. Around Pemberton grew thousands of these trees, attaining their greatest sizes.

To forestry and timber people, Pemberton is perhaps best known as the home of the world's tallest fire lookout tower. Only a short distance from the hotel, on the highest ground, stands the Gloucester Tree. It is a most extraordinary sight. Perched at the top of a karri devoid of all its branches and clad only in clusters of leaves close to its trunk, 200 feet above the ground is what I can only call a crow's nest. When I saw it in 1956, it was perfectly possible to climb, unattended, to the very top and enter the crow's nest to peer down to the forest floor 200 feet below. Spiralling all the way up the clean trunk to the top was an amazing ladder. It was uncanny to climb because every odd rung was a round steel rod driven into the tree and every even rung was in fact a square wood block suspended loose between the steel rods. It was some comfort that all of this was encased in wire netting making a little cage running up and around the tree, but it was certainly not for the faint-hearted or sufferers from vertigo or altitude sickness.

The Gloucester Tree, the world's tallest fire lookout, at Pemberton. Created out of a karri in 1946, it is 200 feet tall and makes an interesting climb.

The next morning I met the local forester and we drove out into the great karri forest to an area where timber was being extracted. Australia has always had a first-class record of good forest husbandry and the forester's job was to monitor and control the licensed felling. It is an adrenalin-raising experience to see and hear a tree more than 200 feet tall, with a trunk 8 feet in diameter at its base and free of branches for over 100 feet, crashing to the ground. These monster faultless logs were then dragged to a log landing before being put onto a log train or logging truck and taken away to a sawmill to be turned into gigantic square cross sections of timber for bridges and other structural work. Karri has the rare combination of being an immensely strong timber available in large sizes free of defects, so the demand is great and the supply limited. When I was there, felling was only of the big mature trees and, where there had been logging in recent years, hundreds of young trees were sprouting up naturally to take advantage of the light from the gap in the forest canopy.

A giant karri falls and a modest feller stands back.

I spent a week in the Pemberton area before moving a little north to Manjimup where the scene was rather different.

Mamjimup did not boast a hotel quite like Pemberton's, and it was not the land of the karri. Instead it claimed to be the centre of an even more esteemed tree and timber, the jarrah. Jarrah is known as a very special wood which played two vital parts in the development of Victorian and Edwardian London. Jarrah's outstanding attribute is its resistance to fire, so for a long period of time it was the greatly preferred, perhaps exclusive, wood for sleepers or cross-ties in the London Underground Railway. Jarrah wood is also very hard-wearing and many of Central London's streets were once paved with end-grain blocks of jarrah since it was the best wearing and least slippery of woods, and preferred to a tarmac surface. I can remember seeing some of these jarrah-paved streets as late as in the 1950s. As if blessed with everything, jarrah is also one of the most long-lasting

Deep in the jarrah forest,
the forestry team works out its next move.

(durable) and strongest of timbers commercially available, so it has always been in demand for heavy construction. Finally jarrah is a rich red-coloured Eucalyptus wood, sometimes highly figured, so selected material makes wonderful panelling and furniture. I still have some jarrah, resawn out of a sleeper taken out of the London Underground at least 50 years ago – the wood is superb.

So, here I was now in the heart of jarrah country at Manjimup. Jarrah covers a much larger area than karri but it is a smaller and less attractive tree – there had to be a drawback somewhere! A mature jarrah reaches less

Jarrah railway sleepers (cross-ties) piled ready for export
in a Western Australian yard in the 1930s.

than half the height of a karri, and the bark is not so smooth but rugged like an oak. It is nonetheless a fine tree.

With the Manjimup forester and his staff I spent another week in and among the jarrahs. We "cruised" over a very large area in a rugged Land Rover which was tested to its limits and never failed. I never could have believed anything like that was possible as we bumped and thumped over cruel terrain. Once we seemed to get lost but we came to no real harm. We didn't exactly camp by a billabong under the shade of a coolabah tree but something very akin to it. This was adventure.

As we drove along we saw kangaroos and emus. A favourite sport I was told was kangaroo racing in the Land Rover. "A good one will go up to 40 miles an hour!" We also saw handsome blackboys, an extraordinary kind of tree. Otherwise known as grass trees, their short fire-blackened trunks are capped by a tussock of grass-like foliage. The best and most mature have several branches. When they flower they produce a tall erect stem on top of the trunk and grass tussock.

All too soon it was time to return to Perth, but it was not without adding more samples to my collection. Back at Perth the good Eaves family welcomed me again, and we had a farewell party before I left on a Super Constellation back to Singapore and Kuala Lumpur.

The grass tree in Western Australia.

I had a goodly haul of slouch hats and real aboriginal boomerangs for friends back at base and stories to tell. But my time was running out and in only three weeks I was notified I was going home.

I had rather hoped I could fly back, both for the alternative experience and for the speed, but it was not to be. Instead "The Management" had stored up a real treat for some of us. It was called S.S. Captain Hobson. I said that Asturias was too good to be a troopship. If it is possible, Captain Hobson was just too bad. Rumour had it that it was actually a converted Burmese coaster. We National Servicemen were trained to rough it, but that deplorable ship actually took emigrants out to New Zealand (convicts deserved a better passage than that!) before bringing servicemen back. Arrival at New Zealand must have been something special. In fact on the next voyage out she broke down and had to be towed 600 miles.

Whereas Asturias, sleek and white, made our journey out in three weeks and her engines made a healthy gentle throbbing noise, Captain Hobson, tubby and black, extended our journey to five weeks and its engines made the most absurd doom-doomp, doom-doomp, doom-doomp noise, at each doomp raising its tone like a question. We passed, at about ten knots, all of the places we had seen on our way out plus a trip to Bombay to change our Indian crew.

Of course as everyone was at least going home for some leave and most were to be demobilized, there was not any pretence of organisation. Dreamy days passed by, punctuated by Tombola games, while a seemingly interminable card-sharping game only ground to a halt when it was simply so hot and sticky at the Red Sea that the players ran out of energy to play. At long, long last, our tropical kit was rejected and we returned to the normal Air Force blue. Finally Captain Hobson chugged into Liverpool and issued the last dooms I was to hear (although I have instant power of recall of that noise).

The rest happened with alarming rapidity – back to a station to hand in the tropical kit, get on a train, and go home.

No prodigal son could have had a greater welcome, but I did not return empty-handed. A day or so after my arrival a large wood box known as a deep sea chest was delivered to the house. It was full of goodies for the family, souvenirs, and the rest of my wood samples. When I left I had 300 species. Now I had 600 and I thought it was a big collection.

Chapter 10
The Growing Collection

I think we all had difficulty adjusting again to civilian life in cool and austere post-war Britain. Life had been so good where we were, but that was all changing, too. Sometimes I wonder if I played a part in bringing independence to the colonies; soon after I moved from Singapore it gained its independence, and the same applied to Malaya which was to become Malaysia. The moment in time had passed. And I had so relished my introduction to Australia that its attraction was unsettling to me.

But time passed, new friendships were made, and it was wonderful to be back with my family again. Gradually things settled down and a family discussion decided that the best thing for me was to gain more timber experience. I was to spend a year with a firm of timber agents in London, and then a year with a company similar to ours in East Anglia, some distance from London. This was all very good experience and greatly helped me in the coming years.

It wasn't long before the wood collecting urge rose to the surface again, if it had ever subsided. There were various samples of woods lying forgotten in corners of the agents' offices and I soon scrounged these. Circumstances dictated that I change the mode of my collecting for some time to come. I was to become an armchair collector; no, not a collector of armchairs but desk-bound.

Somebody, somewhere gave me a copy of the *Empire Forestry Handbook* of 1952. It triggered literally hundreds of letters and harnessed the goodwill and enthusiasm of hundreds of truly delightful people all over the world, nearly all of whom I was never to meet. Not only did it list the addresses of all of the Forestry Research Institutes in what it now conceded was the Commonwealth, but several others elsewhere in the world were listed, too. It was irresistible.

Soon letters were winging their way to New Guinea, Northern and Southern Rhodesia, Sarawak, Nyasaland, Tanganyika, the Somaliland Protectorate, and a host of others. And soon, fortunately, the replies were winging their way back. Virtually all were helpful and few were taken aback at the request. Several countries kept stocks of samples of their timbers in a standard size that was at least as big as mine and others said they would produce them for me. Only one country replied in a rather waspish, superior way that it had better things to do than write to me! Many charged me nothing at all and no charge was anything more than nominal.

Now the postman was calling with ever more exotic letters and parcels, and I was clearly establishing something of a name within the Post Office organisation. One day someone near London posted a parcel of pieces of wood to someone else and forgot to address the parcel, so I received a phone call to ask if they were for me.

British Guiana, British Honduras, British Solomon Isles, Fiji, Jamaica – still the parcels came. Obviously, this ample supply line eventually had to trickle to a halt, but it didn't for quite some time, and, long before it did, another vast avenue opened up before me. Someone told me there was a wood collector's society in America. This amazed me. Up until now I had believed I was on my own. I had not met anyone else who had a hobby and collection like mine. Now it appeared there were many more. All this was apparently the brainchild of Harold Nogle of Port Arthur, Texas. Little did I know that when I sent off a letter to him it was going to trigger correspondence that was to last until he died in 1969. I had a very nice letter back, enclosing a list of samples, many of which I wanted, and some information about the International Wood Collectors Society (IWCS).

Getting dollars in 1957 was not easy and I had to ask for an import permit from the Board of Trade to enable this. This involved a little hyperbole about the important scientific aspect of my work (if there was to be any scientific benefit, it was to come much later). My story came off and I was told, "Well, just this once."

Before too long I received a goodly parcel of several samples, mostly American, filling up a yawning gap from that part of the world. I should have joined the Society then, but my reservation was that I would receive their monthly *Bulletin* which would offer more samples which I would hunger and thirst after, but would not be able to acquire because of the dollar problem. In fact, it was not long after that restrictions were eased, but it was not until 1960 that I eventually joined IWCS.

In those three years, samples were pouring in from research institutes in many countries, but my delay in joining IWCS must have meant that I missed a lot. Once I had signed on the dotted line and the *Bulletins* started flowing in, I opened up a whole new supply line. At the time, membership was quite small and nearly all American. Now there is a large membership worldwide. At the time too, the Society consisted mainly of collectors of wood samples or specimens, often with a scientific interest in wood. Now the members have a wide range of wood interests with a strong element of craftsmanship in wood. This craftsmanship represents the highest skills in woodworking in the world. *World of Wood*, the successor to the *Bulletin*, is a bi-monthly publication for all members, and it contains a huge range of tree and wood topics, ranging from wood technology to wood skill, laced with a general love of wood. The Society also hosts regular meetings throughout America, in Australia, and Britain, where members can share experiences and become very good friends. It is a superb society – for anyone interested in wood for whatever reason, it is a must to join.

Within the Society, I soon established contact, which lasted for many years, with Art Green in Louisiana. Unfortunately, he is no longer with us, but over a long period of time he sourced and supplied samples in a most friendly way. There are 572 species in my collection which he supplied – the largest number from any single source. Altogether 102 members have supplied me with samples over the 48 years that I have been a member. There are too many names to list but I would like to thank all those helpers for all they have done, and continue to do, to build my collection to what it is.

While this great growth was going on within the Society, my pen and typewriter were busy elsewhere, as I kept up the barrage of requests for samples from everywhere I could imagine. I had progress charts of letters sent, letters received, samples on passage, countries contacted and still to be contacted. As I pored over the atlases and my geography approached encyclopedic standards, it occurred to me that a concerted campaign on islands might be a good idea. Obscure islands in particular, I thought, might be pleasantly surprised to have a request for pieces of their woods. In addition, islands are particularly likely to have endemic trees, which in botanical terms means occurring only there.

This campaign was quite remarkably successful, both in terms of new samples and in terms of the immense pleasure it seemed to create not only for me but for those islanders. One of the first I tried was Mauritius which, among other woods, sent me a magnificent piece of their endemic ebony which is the purest black of all the ebonies I have.

Encouraged by my success, I wrote to the nearby French island of Reunion. This certainly qualifies under the obscure category as it is only 38 by 25 miles. Here I was even more successful because they sent me a sample of every species on the island. The Comoro Islands near Madagascar yielded a few more, including their own endemic species of mahogany.

The Seychelles were to be perhaps the most rewarding of all. I had an ancient set of *Encyclopaedia Britannicas* which was a mine of information about the flora and fauna of the islands, so in the case of the Seychelles I especially requested two particular woods. One was the Coco-de-Mer Palm, an endemic palm producing an enormous double coconut with a rather vulgar appearance. Long before these trees were known to have their home in the Seychelles, the fruits had floated onto far distant beaches and it was thought that they were a product of the sea. The other wood I requested was the Bois-de-Fer, or Seychelles ironwood, which I knew was rare.

The very nice reply said the Coco-de-Mer was on its way, with a piece of Bois-de-Fer. "You will notice", they said, "that the Bois-de-Fer specimen is insect damaged and rotted in places. This is because we have no specimens left now and the species is almost extinct. We only discovered a few coppice shoots on an old stump a few weeks ago, which was our first proof that the race had not died out. Nearby lay the remains of a large tree which had fallen a long time ago. From this we managed to select some pieces not too far decayed from which we prepared your specimen." Even today, this is one of my most treasured possessions.

Pitcairn had to be the next island to try, being just about the most remote and unlikely place from which to get wood samples. Beyond the "Mutiny on the Bounty" story, I did not know much about it, but someone had told me it was visited just twice a year by a ship which put the mail onto a small boat for the island before leaving it to its isolation.

Now I know that the island is only 2-1/2 miles by 1 and the population was then perhaps 50, but in my ignorance or perhaps a little guile, I addressed my letter to the Department of Agriculture, Government of Pitcairn. After the requisite year, I received a reply with samples from Elwyn Christian, apparently a descendant of the mutineer Fletcher Christian. There were just four tree types on the island and he had sent them. "Don't bother about paying cash for these," he said, "but if you care to post me a couple of long, strong hair combs I would be very pleased. I have thick dark curly hair. Thanks a lot, Yours sincerely Elwyn Christian." I managed to find some nice steel combs with which he was delighted.

Samoa, Guam, Tonga, the Andaman Isles (renowned for their timbers such as marblewood, padauk, gurjun, and silver-greywood), New Caledonia, Tahiti, Hawaii, the Bahamas, Bermuda, Guadeloupe, Trinidad, St. Lucia, Barbados, Puerto Rico – all yielded their treasures, and the friendly people responded magnificently.

I turned my attention to the Azores, the Canary Islands, and Madeira. Madeira was perhaps the best island of all. The Forest Officer there took to the idea with tremendous enthusiasm and over a period of time collected a sample of every tree type of that verdant isle. We had some wonderful correspondence, and even when he went on holiday to Portugal he asked me if there was anything I wanted from there. This resulted in him charging up a mountain to collect a special oak for me. His interest was postage stamps, so my way of saying thank you was to send him several of these which he wanted.

The challenge of Iron Curtain countries at the height of the Cold War was irresistible. My first success was Hungary where I established a marvelous rapport with a most cultured, intelligent, and helpful man. He could easily have been the brother of my friend in Madeira, for he also charged up mountains and went to off-beat places to collect for me. We corresponded for years. I have a voluminous file of his letters which show just how close he and his family became to me and my family. All this was taking place at a time of high-level confrontation between the Soviet Bloc and the West, so his correspondence was particularly treasured by me and surely risky for him. Photographs, woodprints, even Christmas cards winged their way to us. It was difficult to send any parcels back because the bureaucracy clearly caused him problems, expense, and perhaps risk, however, one day his letter included this:

> "I want to ask you for help for an old friend of mine, a painter, more than 90 years old, living in Transylvania. He is an aquarellist and works with English artist's colours. It is not possible to obtain this colours actually neither in Hungary nor in Roumania. I think this colours are not cheap, so it should be asked how many they cost at first. If they expensive, smaller quantity will also do. Producers are G. Rowney or Winsor & Newton. You should post them direct to Roumania."

Expense or no expense, I was delighted to send them. It was a great thrill not much later to have two very nice water colours from the artist Sandor (Alexander) Pataky, one of the country and one of the town. Two years later in a letter my friend wrote:

> "I regret to inform you that Alexander Pataky died last year at the age of 88. He was the senior of Transylvanian Hungarian painters. Recently he was awarded a Roumanian state prize for artists. His

> paintings are to be found all over the world from Australia to U.S.A. Many of his paintings are in museums. You should keep the paintings he sent you as they are precious."

Precious they certainly are and I have kept and treasured them ever since. Regrettably my Hungarian friend has also now passed away, and I treasure my memories of him, too.

Encouraged by my Hungarian success, I felt the time was ripe to sound out the Great Bear itself, Russia. I wrote off to Moscow. In no time at all, I had a most cordial letter saying that a box of samples was on its way to me. There was no real need for any kind of payment or exchange as it was a pleasure to be of help. Soon after, a nice wood box with a sliding top arrived with my name and address neatly written in Russian and English, and for good measure they called me Professor. The samples were excellent. "We are sorry," they wrote "that we cannot send you the sample of the poplar you asked for, but you could get it by writing to an address in Voronezh south of here where it grows." Again I had a quick reply which said in effect, "We don't know why Moscow asked you to write to us as it is not found here. However, there is a tree growing in the Park in the City and we have been out and cut off a branch for you." Part of the hefty branch arrived soon after. I wondered what sort of response an ordinary Russian would have had at that time by writing to, say, Manchester in the same way.

Letters with exotic stamps and samples continued to arrive in packets, boxes, and even a large jute sack, from Iran, Iraq, Afghanistan, Burundi, Burma, Taiwan, Japan, Mexico, Argentina, Chile, and many more.

But surely the greatest triumph was when, after years of trying, I suddenly had a major breakthrough with China. I had tried several addresses and mostly got no reply, or occasionally my letters were returned "Gone Away" but to what horrible place I did not like to imagine. The breakthrough came when a tree expert had returned from a seed gathering expedition to Mount Omei and he gave me a name and address to write to in Beijing. I wrote post-haste and this man replied suggesting someone else. I feared the worst, but wrote off with a long list and offering samples in exchange. I received a letter typed in perfect English saying that nearly all of the samples I requested were being sent, and they had more. I could not stand the suspense waiting for them to arrive, but at last they did. They were perfect and the Chinese must have gone to an enormous amount of trouble for me because, although they clearly had a standard size which was smaller than mine, in most cases they had prepared samples specially to my standard size. China has a wealth of endemic and intriguing conifers and there were samples of nearly all of these with splendid Chinese labels. My excitement was almost uncontainable. I was delighted to send the requested samples in exchange. We conducted a further exchange so that we each had about 150 new samples. This was some time ago and regrettably the contact does not exist now.

Chapter 11
Innocents Abroad

All of that collecting took place over several years, but meanwhile I had finished my two years of timber trade training. It was 1958 and I was ready to start with the family business. I was given a dark blue Ford Anglia car and a price list. I was to be a sales representative to open up new territory south-east of London. How different things were in those days! Despite my very good grounding, I had had no training in doing anything like that and there was no such thing as trade telephone directories. But it worked, and it worked quite well. A good deal meant not only a satisfied customer, but I found it very satisfactory, too. By now most of the post-war problems were out of the way, softwood was in plentiful supply again, and those mysterious hardwoods had disappeared from our yard. Houses were being built, and timber was easy for us to buy and sell. I was meeting craftsmen, and it was a joy to watch their work. I was finding that I could sell and that hard work brought rewards. It was nice at last to put some of my training into practice.

While I had been training, I had had little spare cash that wasn't spent on writing off for samples, so the prospect of foreign travel was not open to me. But now I had a car and some friends, so the idea of combining a little continental travel with some good wood hunting became a reality.

Teaming up with three former colleagues from London, in 1959 we decided upon taking the car down through Belgium, Germany, and Switzerland to the Italian Lakes, and returning via France. As one friend was now working in the wine trade and had contacts at strategic points, it sounded even nicer. It was an all-round great trip. The scenery was great, the drinking was great, and the banter was great. I realise now that I did not then have much idea about trees and timbers of Central Europe, but I decided to use the opportunities to see if I could add to my collection. In Munich, I bought a book about German trees which would have been of limited use since it was in German and I knew none. But, it had beautifully painted pictures of all of the leaves, flowers, and fruits, and it proved invaluable.

When we returned, there was a modest pile of small logs awaiting conversion to wood samples.

We had such a pleasant holiday that we decided to team up again the next year, 1960, but this time we set our sights southward to Spain. We went down the Rhone Valley and entered Spain at the north-east extremity, south of Perpignan.

It is difficult to imagine how different the Spain of 1960 was from the Spain of today. Franco's regime had the country in an iron Castro grip. It had been forced to turn in on itself, so it had little idea of, or interest in, the outside world. It was straight-laced, delightfully behind the times, and unspoiled but very poor. The Spanish atmosphere was exactly like Greece's under the repressive regime of the Colonels some years later. But it was such a beautiful place with an air of drama and romance that I fell for her charms hook, line, and sinker.

We moved down the coast looking in at the landmark towns, but Spain was uncharacteristically wet. This time I knew already that I did not have much idea about the trees and timbers of Spain, but I still thought I would like to add some more samples. In Barcelona, I bought a book but this time I was not so lucky. It was in the Catalan language and only had rather unhelpful line drawings. It soon became clear that I needed to teach myself a lot more about tree classification, and that Spain has quite a large number of different trees (in fact more than any other European country except Greece).

As we continued south from Barcelona, the weather perked up and brought that searing Spanish heat I learned to love. The olive trees were roasting, mesmerised by the sun, and the peaches, figs, and apricots were coming exquisitely to their full juicy ripeness. In the matorral or maquis, the pervading rich aroma of rosemary and sage sweetened everything, and the Cistus bushes covered hillsides with their sticky leaves and ephemeral showy flowers. Cicadas chirrupped endlessly, while here and there an unusual bright green lizard basked in the sun or scurried away at a footfall. Only man was busy picking fruit or loading a cart or coaxing a mule – or driving restlessly south.

When we arrived at Sitges with its palm-clad sea front, it seemed just perfect. As with so many places, I have never been back, so I wonder what it looks like now. Then it had not sold its soul to the tourist and every inch of it radiated Spain. We found one of those nice old-fashioned hotels with shoeshine boys, porters, and plump but genial receptionists. We also found a nice little restaurant where we made our first acquaintance with

Paella Valenciana in such bounteous amounts as to call for lubrication with bottle after bottle of vino rosado at a minimal price per bottle. Late at night, heavy of step, we returned to the hotel wondering at the resilience of the human digestive system.

When I awoke at 6:00 the next morning, there seemed to be a premature buzz in the air and I lamented the bacchanalia of the night before. But there really was a buzz outside. Surely even the tough Spaniards are not afoot as early as this? I stumbled to the window overlooking the street. There were clusters of people all along it, and at intervals men were drawing strange patterns on the road with chalk. Soon we were all up on our feet again and wondering what on earth was going on. Downstairs at breakfast of rolls and butter, rich honey, and soothing black coffee, we asked the genial waiter.

> "Ah Señores, today is the Festival of Corpus Christi. Today you will see all the streets of Sitges turned into carpets of flowers. Every street will be decorated with a different pattern. All the people of the street will work together to make their street carpet the most beautiful – like a Turkish carpet. For the winner there will be a big prize. Then in the evening when the heat at last cools down, the religious procession will walk through the town and trample all the carpets and break them all up. It is necessary so they do not

Carpet of flowers and leaves in a street at Sitges, Spain, for the festival of Corpus Christi, 1960.

fade away and it is the Festival of Corpus Christi which we all celebrate, Ave Maria. Long before you were awake the people have gone out into the country collecting flowers and leaves and soon you will see the beginning of the carpets. Buenos Dias, y Gracias, Señores."

Sure enough, when we went out into the streets the carpets were already taking shape and the patterns had been marked out fully. Some streets were carpeted just with flowers, others with flowers and leaves, and some with leaves of different colours with only a few flowers for variation. One wondered whether the surrounding country- side now looked as if a plague of locusts had struck. The effect was stunning. Jubilant at the fruits of their labours, the workers now vanished back into the cool of their patios for the siesta while tourists swam in the sea.

As the evening drew on and the heat again began to abate, the palpable festive aura developed again. Crowds lined the streets and the procession started in the square where a huge floral cross, flanked with bunches of Bougainvillea, had been created. It snaked through all of the streets, spreading the fading flowers far and wide, creating bonanza employment for the sweepers, and then out onto the sea front, amid a carnival mixture of religion and festive balloons. We stood watching unobtrusively.

Dress at that repressive Catholic-Puritan Franco time was strictly controlled – no shorts could be worn by men or women and all sleeves had to be long. I felt an old fishing hat would not be an inappropriate thing to wear in the still-warm sunshine, but I miscalculated. One of the ubiquitous Civil Guards wearing his glazed-leather-three-cornered hat clearly disagreed. He came up to me and remonstrated in unintelligible Spanish. I could not understand even the gist of it. He held his rifle up to my head and tipped the hat off it. Now I understood.

We headed on down to Valencia and beyond, and then cut back inland to Saragossa while the sun blazed down in unstinted exuberance. The south slopes were relatively dry, but as soon as we crossed over the Pyrenees into France, the difference was palpable. Here were rushing torrents of water, the shade of lush trees, and the cool of further north. My knowledge of trees had been improving despite limited resources, and when we got back home there were a few more treasures for my collection.

~ ~ ~ ~ ~ ~ ~ ~ ~ ~ ~ ~

I met Susan at a staff dance party at a school for blind children which she supervised.

My friends and I had formed the excellent habit of meeting at a pub on Saturday mornings, for a jolly pint or so and to allow plans to develop for an even more jolly evening. Nothing was ever planned in advance. Somebody always knew somewhere where there was some kind of jollification. On one Saturday morning, the news was that there was to be a party at the Blind Children's School. So we went. I had only been there a few minutes when I noticed Susan. "That's the girl for me," I decided before I had even spoken to her. It was a whirlwind courtship and in six months we were married.

There was no particular reason for choosing the island of Mallorca for our honeymoon in that year, 1961. Nor were there any devious, ulterior sample-hunting motives – it just happened that way.

We had one of those lovely, old-fashioned weddings. It was August, the weather was fine, and we married in a village church with a large congregation of our friends and relations dressed in full regalia. I had not told Susan where the honeymoon was to be. She only knew it was abroad. She had not been abroad before, and had never flown, so when we arrived at Palma and the plane door opened, the inrush of sweltering heat staggered her. One could not say that Mallorca was unspoiled in 1961, but it has developed a lot since then. The outskirts of Palma were already well-developed, but by hiring a car we could soon drive off into traditional farming country or up into the Mountains in the north-west. It just happened that I knew a lot more about Mediterranean trees and that we did not come home empty-handed. It was a glorious honeymoon. The sun shone without a break, the scenery was gorgeous, the food delicious, and the atmosphere relaxing. We reveled in seeing the small holdings growing oranges, lemons, figs, grapes, melons, apricots, peaches, olives, and carobs. When we saw one these small holdings for sale for a tiny sum, we were tempted to buy and escape the gloomy north. How I wish we had.

It was only when we landed back at London that Susan revealed that she had been terrified of the flights and said she would never fly again – and she has not. Nothing untoward happened on the flights, nor did she worry about the more adrenalin-pumping parts like taking off and landing. It was just being suspended in mid air that worried her. I thoroughly respected this, even if flying for me has exactly the opposite effect. I revel in it and, the more interesting it is, the more my metabolism seems to rise to the occasion. Also I think I got thoroughly hooked in Malaya, but it did not stop a deep compassion that it was such an awful experience for Susan. Fortunately, she is quite happy for me to go on my expeditions.

Chapter 12
New Directions

While all these otherwise happy marital events were taking place, events were warming up to another of those turning points that seem to have shaped my life.

I was enjoying my job as sales representative for the family business and it was going well. I was meeting many nice people, selling was fun, and my wander lust could be satisfied every day.

But you will recall, I am sure, the uneasy alliance between members of our family whose religious thoughts, if any at all, were superficial and relaxed, as against the ardour of the Plymouth Brethren members. Religion was not discussed, and by concentrating on other subjects it was possible to co-exist. My Uncle Alec and his son Edmund were jointly running the business with a non-family member and it was all going well.

Quite suddenly a crisis brewed up. It seems that for some time an intensification of religious fervour had penetrated the Plymouth Brethren movement. In America, a zealot by the name of James Taylor had gained control and was tightening the religious screw. The effect seemed to drive a wedge between the more relaxed brethren, who basically wanted to opt out of the religious strictures, and those whose fervour conditioned them to accept, and indeed thrive on, ever tightening religious strictures and isolationism. The former became known as the Open Brethren and the latter the Exclusive Brethren. If both husband and wife were not of like belief, they were to separate. No brother or believer was to take a meal with a non-believer (i.e., anyone else). Professional people such as doctors and lawyers were not to be members of any professional bodies that were not exclusively brethren. It was not permitted to have television in Exclusive households, and one of the most bizarre rules was that neckties could not be worn.

I knew nothing of all this until one day my cousin Edmund said, "Richard, we are going to have to sell the business." The Exclusive family members had found themselves unable to continue to associate themselves with the rest of the family and their regime required them to take their money out. Unfortunately, it seems that neither family faction had sufficient funds to buy out the other and it may well be that the will was not there for it anyway. Whatever the background, the decision was taken – the business, after trading 165 years, was to be sold. Worse than that, it was to be sold off in four constituent parts and not as a whole, for reasons I never learned. There would be no job for me, and the family heritage for which I had worked so hard and long was not to be passed on to me. I was rocked back onto my heels and utterly devastated. It was like losing the family manorial estate. If it had been the result of some wild thoroughly disreputable gambling debt by a reprobate member of the family, I think it would have been more digestible than for it to disappear in this sterile way. I thought of my ancestors' aggregate efforts and could have cried with despair.

The Company was in good financial shape, so it seems amazing that it was not possible to find one buyer for it as a whole. Instead, the wharf was sold to one company, and perhaps it is some comfort that a small part is still a timber yard. The Company and telephone number and so on were sold off to timber neighbours and the few remaining staff went with them. A subsidiary in Southend some distance away was sold to yet another timber company which has since closed. As for the original Benjamin Street in Clerkenwell, I hate to have to relate that the site was sold, and the fine old building pulled down to be replaced by sad, characterless apartments. What a pity that we were not so conservation-minded in those days. That historic building, the last remaining timber yard in the old City of London, should have been preserved with its unique framing of giant baulks of timber.

But life goes on and I was only at the beginning of my working career. At the time, I regretted that I had so single-mindedly dedicated myself to wood. (I certainly do not have that regret now – wood has served me well.) We had only recently married when this debacle was acted out. I decided to advertise in the *Timber Trades Journal* Situations Wanted column saying that, because of the closure of our company, I was looking for a position. Such was the buoyancy of the timber trade in the 1960s that I received 16 job offers. The most persistent wooer, based near Bristol (150 miles away), sent me a telegram and said I was not to consider anything until I had been to see him. Within two weeks, I started my new job of selling in the Bristol area with a view to creating a new branch there. I was to pioneer that new branch.

We could just about afford a lovely old house in a village near Bristol. Situated next to the church, it had belonged to the village builder, which should have been a good recommendation but it was in need of some attention. We had no money beyond my monthly income, and little to no experience in restoring old houses, but timber was readily obtainable at favourable prices and I thought I knew all about carpentry. I was quickly to learn that making free-standing objects is a very different story from affixing shelves or cupboards to crazily leaning thick walls of soft mortar. I remember the loss of face to my dear wife when I set about this work with a confident air only to find that my nice pieces of straight wood leant wildly and refused to fit together when I "offered them up" (what a strange, ritualistic phrase that is for builders to use) to the recalcitrant walls. Few of the handy power tools that we take for granted now were available then, and even if they had been we could not have afforded them. As I punched holes in the wall to take plugs, the nice small hole grew and grew in size as the soft mortar broke up. I was in despair and panic. Susan had the same learning pains with wallpaper which would not adapt to crazy angles. But we learned by trial and error. One of our few luxuries was a television, but even then we could only buy a series of geriatric sets which gave us a few months of deteriorating viewing.

Gradually we learned how to do all the clever things. The mysteries of plastering and plumbing were learned after some disastrous first attempts. I came home from work one evening to find Susan had demolished a hideous tiled fireplace, revealing a gaping hole and leaving debris and dust over a large area. I felt like putting in some overtime at work, but as the dust settled I realised she had exposed a very fine old fireplace. One Saturday afternoon found us hacking away at what seemed like acres of old tarmac-surfaced ground, to create lawns and flower beds. I spent the housekeeping money to buy a load of earth and soon we were beginning to look respectable.

Our financial struggles restricted but did not halt the growth of the wood collection. Much of the armchair collecting took place at this time. My island hunting time had largely passed, but now by chance I had a new lead. Someone had told me about the United Nations Food and Agriculture Organization's *Directory of Forest Products Research Institutions*. With uncanny skill, fate was stepping in yet again to point me in new directions. Now I had access to organisations all over the world, and soon the pen and loyal, ancient typewriter (another family heirloom dating back to 1927) were pressed into service again and more samples continued to arrive.

Gradually, our finances improved. The job was going well and inflation was then a benefit to us as salaries rose and mortgage payments became comparatively smaller. In 1964, Emma, the oldest of our three daughters, was born and, in 1965, we decided we could afford a proper holiday again. And with a bit of luck, the collection could have an uplift too.

Chapter 13
Vistas de España

One bonus of my work visiting builders around Bristol was that frequently they were building houses in old or overgrown gardens. Trees and shrubs were to be sacrificed in the process and this gave me a valuable opportunity to add to the collection. Frequently, my lack of experience meant that I could not identify the trees. I found Bristol University very helpful in this, and in the course of one visit I was introduced to their library. Here I found many books about plants and trees, and before long I honed in on an excellent, if old, book about the plants of a country I had already fallen in love with, Spain. Sifting through its many pages to find the trees, it became apparent that there were several special species to hunt for in southern Spain, and there was the mystical city of Granada.

Part of the Court of the Lions in the Alhambra Palace, Granada, Spain.

By a happy chancc, I had some books about Granada and especially its romantic and dramatic history, first with the idyllic Alhambra Palace as seat of government of the Moors and then the saga of its capture and the exclusion of the Moors from Spain. William Prescott's classic, *History of the Reign of Ferdinand and Isabella*, written in 1837, and Washington Irving's dreamy, *The Alhambra*, written in 1896, are enough to kindle the imagination and inform about Granada. Even if the greatest of travelers in Spain, George Borrow, did not include his experiences among the gypsies of Granada in his book, *The Bible in Spain*, written in 1842, his racy, melodramatic descriptions of travel there in the 1830s are enough to kindle the enthusiasm of the coldest heart. *The Bible in Spain* is surely the most uninviting name for one of the best travel books ever written. True, Borrow was working for the British and Foreign Bible Society and was traveling to Spain to distribute bibles in Spanish in defiance of a deeply entrenched priestly hierarchy, but the book is really about his travel on horseback over a large area of Spain ravaged by civil war, accompanied only by a guide, and passing through wild scenery infested with "the factious" and bandidos. He mixes with gypsies and humble peasants, eschewing the prosperous and pretentious. For the presumption of selling Bibles in Spanish for almost nothing, he is locked up in Madrid's prison for a short time. But this only added to his reservoir of experience and defiance. The book was banned in Franco's time and would probably be difficult to obtain now, but I doubt there has ever been a better travel story.

Our plan to visit Spain in 1965 was now more than just a pipe dream. We had information from the Tourist Office in Granada and listened to a course of basic Spanish lessons on the radio, so at last in July we were ready to go. After crossing the channel, our plan was to drive through France through the night and be in Spain the next evening. At that time Susan was not a driver, so it all rested with me. We left Dieppe at midnight and after about two hours of uneventful driving I began to yawn, so we decided to stop by the roadside. It was totally flat, pitch-dark, and windless. We were nowhere near the great battlefields, but we had recently seen Great War programmes on television and thoughts of sombre gravestones churned in Susan's mind. There was nothing for it; we had to go on. We had croissants for breakfast at Poitiers, and for most of the rest of the day we continued south on that nearly perfectly straight road that takes you 150 miles from Angouleme to Biarritz, skirting Bordeaux. At eight o'clock, we finally pulled into San Sebastian after 600 miles. Exhausted, we booked a room at a modest but friendly hotel, going through the requisite formalities of filling in voluminous forms and depositing my passport.

Thoroughly refreshed the next morning, we set off to cover the 300 miles to Madrid. Now Spain put on its best air of welcome for us. The sun rose gallantly in the sky and sent waves of shimmering mirage heat. The ripening corn bent gently in response to a warming southerly wind. The villages basked, perched on their hilltops. The only restraint to pleasure was the slight menace of two armed guards at the approach to each village.

We had covered 200 miles, things were going well, and life was really very pleasant as we drove into the siesta-bound town of Aranda de Duero,

> "Of course you've got your passport" said Susan.
>
> "Oh my God, I've left it at San Sebastian."
>
> "It'll add 400 miles to go back" (and mess up the carefully planned itinerary).
>
> "Look, that looks as though it may be the Town Hall of Aranda. Let's call in there and ask if they can ring the hotel in San Sebastian and ask them to send it on to us."

It was at this point that my newly learned Spanish could be tested and was found wanting. What I had thought was the town hall was in fact the savings bank, and their interest was in Spanish deposits rather than passportless Englishmen.

> "Go to the Civil Guard barracks over there and ask them," they said.

It is difficult to portray the repressive atmosphere in the Spain of Franco's time. Englishmen without passports were surely to be of the greatest interest to them.

In the heat of the mid-day sun, the parade ground was deserted. Only a few trucks and jeeps simmered in the heat. In the wood shed the guards reposed, the smoke from their Caporal cigarillos drifting lazily out of their slightly parted lips.

> "Si, Señor?"
>
> "I, ah, I have left my passport in San Sebastian in a hotel," I said, fingering my phrasebook, "we are, ah, driving on to Madrid tonight and then to Granada tomorrow. Please can you help us to ring the hotel in San Sebastian (fortunately I remembered its name), so that I can ask them to forward it to us in Granada; I will pay the phone bill."

I waited for the springing to action from the torpor and the jingling of handcuffs. There was no movement.

> "Oh, you have left your passport in San Sebastian? We shall need a conferencia; it will take two hours. You must wait."
>
> "A conference?"
>
> "If you refer to your phrasebook, Señor, you will see that a conferencia is a long-distance telephone call. You may wait in your car."

Torpor returned to the smoke-filled room. An hour passed and a splendid Mercedes with venetian blinds drove in. It contained the Big Chief. The guards sprang to attention and saluted. The gruff, well-fed chief swaggered and grunted. The guards postured in sycophancy.

The chief drove off, and they lolled back again. The heat soared and two hours had passed.

> "Ah, how is the conferencia?"

Three hours had passed.

> "I think, Señor, it will be better if you drive on to Madrid. We will speak to the hotel and ask them to send your passport on to the British Consul in Granada."
>
> "Oh, thank you very much, that is very kind. Ah, is there a British Consul in Granada?"
>
> "We don't know; we are not sure."
>
> "Well, I know there is a Tourist Office in Granada, so could you please ask them to send it on to there. Here is the address."
>
> "Si, Señor, de nada, Señor. Adios y buen viaje."

We took to the road again, slightly, but only slightly, relieved, for what will happen when we get to Madrid? Fortunately, we were not asked for any passport there and the next day we set off for the 300 miles to Granada. As the sun began its unrelenting mission of baking heat, we crossed La Mancha of Don Quixote and the windmills fame, and up into the Sierra Morena to once-notorious Despeñaperros where Borrow, as always, just missed being robbed or worse by bandidos even though he heard strange calls and screams from the menacing crags.

Our approach to Granada was every bit as dramatic and exciting as we found our way up the Cuesta Gomerez, the narrow and steep street which leads to the Alhambra. Just at the entrance, the Puerta de las Granadas, we stayed at the Pension America.

After breakfast the next morning I walked down to the bottom of the Cuesta to the Tourist Office.

> "I wonder if by any chance you have my passport from San Sebastian?"

The *Oxford English Dictionary* describes a slob as a "dull slow untidy person, a careless workman." I would add insolent and fat.

> "Pasaporte? This is not pasaporte office, this is tourist office."

"I know, but..."

"No, we have not your pasaporte, we have tourist leaflets."

Perhaps, I thought, it would be better to make a strategic withdrawal and try again the next day.

We spent a glorious day at the Alhambra and the next morning I braved all at the Tourist Office, where the mood had turned more aggressive. "Señor, I tell you this is Tourist Office; we have not your pasaporte."

Back at the Pension, it occurred to me to ask whether there was in fact a British Consul in Granada, and sure enough there was. "Of course," I said to myself convincingly, "they have sent it there after all."

It was half past ten by the time I found the Consul's house. It was large and Victorian, situated near the Alhambra. I knocked at the door repeatedly.

"What do you want?" a woman's wizened face appeared at a window.

"I am British." I said, seeing this as a passport in itself. "I wonder if you may have received my passport sent down from San Sebastian?"

"You had better come in. I'll get my brother; he's the Consul."

I went into their sitting room which was lined with photographs of polo teams of the British Raj in India. He came down in his dressing gown but without a hearing aid which would have helped.

"He's lost his passport, dear."

"Well, it's no good coming here; he'll have to see the Consul in Seville. I don't deal with passports here."

"But he's not going to Seville, dear."

"He'll have to. Lovely place Granada, isn't it? The Alhambra, the Generalife, the Cathedral. Have you been up the Sierra Nevada?"

I returned to the Pension with mixed feelings – mixed panic and despair. I then thought to ask at the desk if they could possibly have a conferencia with the hotel at San Sebastian to see if any light could be thrown on the gathering gloom.

"You are going out for the day, Señor y Señora? Have a nice time and leave it to us; we will ring the hotel."

We had a nice time and when we got back the receptionist was all smiles.

"Your passport, Señor, is at the Tourist Office. If you go down tomorrow morning they will give it to you."

It was not easy to suppress my feelings of trauma and to project expressions of deep gratitude.

The next morning I crept into the Tourist Office hoping to slink past the slob. He beamed magnanimously at me.

"Ah, Señor, your pasaporte here. I told you it would come."

While all these Spanish pleasantries had been going on, we had cast off care and had two splendid days. I have since returned three times to the Alhambra, and it has never lost its charm. Externally, weighing heavily on its crag overlooking the City, it is a rugged fortress softened only by the backdrop of the snows of the Sierra Nevada, but within it is pure oriental fantasy – the Moorish Courts of the Myrtles and the Lions flanked by tent-like colonnades of delicate pillars and everywhere the amply flowing runnels, the blessing of water for settlers from Africa. Above the palace are the Generalife Gardens, lined with majestic cypress trees; above them is the Silla del Moro, the "seat of the Moor" to look out over Granada, and above that, hills and the eternal snows.

The second day was devoted to botanical exploration. We drove out of Granada and up a mountain road. At 7,000 feet on a very steep slope overlooking a river, we stopped for our picnic lunch.

"I shall only be two hours," I said to Susan, "half an hour to drop down the 2,000 feet to the river, half an hour looking around, and an hour to climb back."

"That's all right; I have my sketch pad and the scenery's superb. Just leave a note in Spanish in case anyone comes to talk to me."

I penned a note: "My husband has gone down the valley. He will be back in two hours. I am well."

And then I set off down the steep south-facing slope. My plans, rehearsed over many months, began to go awry. The mountainside seemed much steeper than I had imagined, and the surface was loose gravel and with large awkward rocks. Even at 7,000 feet the sun beat down mercilessly. It took an hour to descend, but I reached the river without mishap and crossed it quite easily. The north-facing slope was covered with lush woodland. An hour had passed but, never mind, one more hour and I would clamber back to Susan who would be so enjoying sketching that she would not notice that I was an hour late. I had not reckoned on the ardour of the climb back. The sun shone down with greater vigour, the slope seemed even steeper, the rocks more awkward, and the gravel looser. My rucksack was heavy and I was panting with thirst. It took me three hours of grueling climbing before I finally was able to overcome the last obstacle to reach the road. Emotions were mixed – Susan was just about delirious and I was full of contrition.

One of the species-rich valleys in the Sierra Nevada, Spain, from 7,000 feet. The valley bottom is a steep 2,000 feet below.

We left Granada with reluctance and moved south to the coast which at that time was virtually undeveloped east of Malaga. We stayed there for three days. One morning we woke up to find the car covered with red dust blown over from the Sahara. The heat pounded down as we visited dramatic Ronda and built up as we neared Madrid on our way back. The temperature climbed to 105°F. We stuck to the seats, and opening windows made things worse, for it was like opening a furnace door. The wind from the Sahara had risen to a searing crescendo.

When we returned to England, Susan had added two other difficulties of southern sampling expeditions to her fear of flying. She simply could not take the heat and she did not like to be where people did not speak English so that she could understand. Of spending five-hour sessions on mountainsides she said nothing.

Chapter 14
Southern Comfort

A year or two passed before I could again succumb to the caressing embrace of the warm south.

The family was enlarged by the arrival of Hannah in 1966 and Josephine in 1969. The wood collection continued to grow too, even if calls on finances made me again an armchair collector. At length the call could not be resisted any more. Sicily and southern Italy were now beckoning. Research had told me that they promised interesting species. I flew down to Palermo, that anarchic mafioso city. In a burst of recklessness, I hired a car and drove through its centre. In no other city have I driven amid such total disregard for the rules of the road. Traffic lights and point-duty policemen were a wasted investment. The only time I saw any driver take any notice of a point-duty policeman was when his girlfriend rushed up to kiss him, bringing everyone to a halt and occasioning a round of cheers and applause.

Up in the mountains I was successful in finding some more new species. Sicily has some splendid ancient remains. Lovers of Greek temples can see seven of them well-preserved in a cluster at Agrigento. Syracuse bristles with catacombs and other ruins; Taormina has its magnificent seaside Roman theatre.

Before I left Sicily, I was determined to see all I could of Mount Etna. I drove through the city of Catania and up the southern slope. The circumference of Etna, at base, is over 100 miles and it rises to 11,000 feet. The lower slopes are sub-tropical cultivation and above this is dense chestnut forest including one hollow tree which is said once to have sheltered 100 horsemen in a storm. In early summer it is a tree-lover's paradise, for the Mount Etna Broom is in full flower and its delicious fragrance pervades everything. In the rich soil, it is no mere man-high shrub but a luxuriant tree up to 30 feet tall. At 6,000 feet is the Refugio de Sapienza, which I reached by an easy road. Here I left the car and continued by cable car. The last part was done by Land Rover. The scene was like some enormous reek-

Beside the awful crater at the top of Mount Etna, 1972.

ing slag heap. There was an overwhelming sulphur stench and the clinkers were huge. At the top were three craters. I walked to the nearest one and peered over its sheer edge. Below was a grumbling, rumbling, steaming smoking mass. Every so often lumps of molten lava would be lobbed, hissing, out of its horrible maw, and they could land around you so you had to be ready to move. I waited for one to cool and picked it up. I still have it. I stayed the night in the Sapienza refuge. It was uncanny to be sleeping comfortably on the edge of such a hostile mountain, with great lava flows all around me. I am not sure whether the refuge is still there. There have been some fiendish eruptions since.

I crossed over to Italy and drove up to Naples, not missing Pompeii which no one should do, nor Paestum where rows of Greek temples in a wonderful state of preservation suddenly seem to materialise out of corn fields.

Greece could not be far behind and it was not. In the first of three very successful expeditions, I landed among the regime of the Colonels. If I thought Franco's hold on Spain was draconian, I was to learn that the land of the Draco was to be home of an even more repressive regime. "Long Live the 21st April 1967" and "Long Live the Army" were slogans suspended on scaffolding at every entrance to every town. Soldiers were in evidence everywhere and people were far too scared to say anything to me. I suppose I was rather foolish to be hiring a car alone with only a phrasebook and to drive it from one end of the country to the other, nudging the Bulgarian

Nearing the top of Mount Chelmos in Peloponnese, Greece, with a Greek guide who had hidden there as a partisan fighter during World War II.

and Turkish borders at a time of border stress. But I came to no harm. Indeed, the worst experience I had was when I had stopped the car in the mountains for a little light relief. It was an extraordinary thing that wherever I parked the car in Greece, someone appeared out of nowhere. I was only a short distance from the car which I had locked, but I had carelessly left my wallet on the front seat. Sure enough, a herdsman appeared from nowhere and saw the wallet. He tried the doors. I thought he would go away, but he picked up a large stone ready to smash a window. At this point, I felt valour was the better part of discretion and emerged from my camouflage uttering a volley of the most vulgar Anglo-Saxon invective. This seemed to throw him and I regained the initiative. Even if, before long, he volunteered a similar tirade of Greek abuse, he seemed to have lost his resolve and, satisfied with a few cigarettes, he melted away as mysteriously as he had appeared.

The Acropolis, Olympia, Mycenae, Tiryns, and Delphi. Surely no nucleus of religion anywhere in the world can have such a compelling atmosphere as Delphi, the navel of the world, etched into its mountain side?

On other mountains, Greece revealed more treasures and the collection continued to grow. On my second expedition I covered the Peloponnese, and my information led me to the village of Kalavryta in the north, near Mount Chelmos. What a story it has to tell. In the War, men of Kalavryta had taken to the hills as partisan fighters and were harrying the

German troops in occupation. One night they ambushed a group of soldiers and killed them all except one (there is always one) who managed to struggle back to base to report what had happened. The next day the Germans came into Kalavryta, took all the males over 13, and shot all 1,300 of them. Then they set fire to the village and burnt it to the ground. They even went up to the monastery perched on a precipice over the village and threw all of the monks over the edge. The women of Kalavryta moved to the next village until the end of the war, when new husbands were found, Kalavryta was rebuilt and repopulated. Nearby is a large, poignant war memorial.

Kalavryta is happy now, and when I requested a guide to take me up Mount Chelmos I found a keen volunteer in one of the partisan survivors who, although in his 70s, bounded up the mountains while singing happy partisan songs. Back in the village we had some convivial times and his son corresponded with me for several years.

My third odyssey was to Crete where, in the west, away from the tourist concentrations, I met other extremely friendly Greeks. I remember driving up into the mountains on a hair-raising road from a deserted Frankish castle on a beach. In the mid-day sun, I arrived in a very nice little mountain village and stopped at its taverna for some refreshment. As usually seemed to happen, an English speaker appeared from nowhere and more and more villagers arrived for their mid-day tipple. As each arrived he insisted on buying everyone a powerful glass of raki. A man who had been picking loquat fruits in the valley came up and deposited a few of them at each table before he drank. I still have a tree which I raised from one of the seeds.

I was pleased to buy my round for them, and as the numbers grew and the raki flowed I asked the English speaker who these various people were.

> "Oh, he is shopkeeper, he is baker, he is farmer, he is policeman."
>
> "Policeman? But I have to get in that car and drive away soon!"
>
> "Don't worry – it is quite all right" he said.

Until it's time to go, I thought. At length with fond farewells I made for the car to drive down those deserted but hairpin mountain roads. Now comes the punchline, I thought, but no.

> "Goodbye, Goodbye." They waved amiably.

In Heraklion there is a picturesque street, then given over entirely to butchers' shops. Each butcher's chopping block was a large log of cypress or eucalyptus wood mounted on three legs like a stool. Of course with the passage of time, the tops would become eroded. There was a man who made his living touring butchers' shops re-flattening these blocks. He was a consummate craftsman. He had three adzes (an ancient tool like an axe but with a blade positioned like a mattock), and he spent some of his time honing these to razor sharpness. Then as crowds sauntered by, he chipped casually away at blocks with his murderous weapons until the tops were so flat that not a chip mark could be seen. It was spellbinding to watch.

The last of these sagas was a business trip with many others to study and tour Canada's timber industry. We visited Montreal, Ottawa, and Calgary. At Banff, we took that marvelous train through the Rockies to Vancouver. With its three engines and countless carriages, not only the Rockies were a great sight. To sit in an observation dome, sipping a gin and tonic while craning one's neck to see the mountains above was a very civilised pleasure, and Vancouver seemed to me to be one of the most beautifully situated and pulsating cities I had seen.

This was to be the last of such visits for many years. Other demands of time and money, including raising our family, were squeezing them out.

My job had taken us further west to the county of Cornwall. Sadly, my mother had died, though she left us the means to move to another house. I now had the chance to get my wood workshop set up and my samples properly housed.

Chapter 15
Round and About the Collection

Wood samples reach me in a variety of forms, whether collected by me (certainly the minority) or sent from afar. They may be in log form, which have varied from purpose-cut small logs about six inches long and four inches in diameter to portions of large trunks saved especially for me after a clearance by a landowner. They may be half logs or pieces sawn or hewn out of logs, with edges "waney" or rounded in log form. They may be fully sawn oversize but absolutely green, or they may be fragments from treasured pieces decades old and bone-dry. They may be sawn, planed, or sanded on one or more face. They may be precision finished to a standard size and painstakingly labeled, or there may be a scrawl on them with just enough information for identification.

Latin botanical names are an absolutely essential criterion for my samples. Botanists play around with Latin names for plants, but they are using a universal and dead, unchanging language so it is usually possible to be sure of what is being referenced in that Latin name. Trade and local names are totally unreliable. Even in Europe, oak is also ek, eich, chêne, roble, encina, farnia, alcornoque, and a host of others. In Africa, one species may have hundreds of tribal names. Conversely, many different woods may be given one name, so-called oaks, mahoganies, walnuts, pines, and ironwoods abound all over the world. Some are "true" woods, but many are imposter names coined for convenience or commercial gain. Others such as ironwood are used to describe any heavy hard wood, and the name may apply to a host of unrelated woods.

At the time of writing this, I have recently passed a total of 7,000 samples, each of a different species. No wood enters my collection unless I am satisfied of its validity as a botanical species. If it arrives in large log form, a piece will be split out and trimmed oversize with an axe. A small log will be sawn to an oversize piece. This must be done quickly to prevent splitting and decay. Because wood dries most quickly from the ends of logs, planks, or boards, it is usually necessary to paint these ends with liquid wax to slow the drying and prevent end-splitting. There is quite an art in knowing how much oversize to cut the sample in its wet state to ensure that there will be enough left to produce the dry one. It must not be too big so that there is a risk of splitting. Some woods, especially conifers, are generally quite stable, but others, such as some eucalypts, will "move" mercilessly or split disastrously as they dry out. If you have not allowed for this, your fine sample will become valueless before your very eyes as it warps beyond recovery.

Once they have dried over several weeks, the samples are ready for a final fine-planing to my standard size, stamped with the accession number on both ends, labeled, catalogued, and entered into the collection. A few are so hard or the grain so "wild" that they cannot be planed satisfactorily and then they have to be sawn a little oversize and sanded down to standard. Otherwise they are not further finished in any way, since I like to see the wood presented as its natural best.

Working with wood is one of the joys of life. Every new piece is a new experience. Until the saw makes its first bite or the plane its first cut, one cannot predict the reaction even after years of trying new woods. Some pines and mahoganies plane like silk. The plane slides smoothly over them and the hidden beauty or sheen is brought to the surface. Others are as fiercely resistant as they are so intensely hard, greasy, and straight-grained that the plane slides ineffectually over them no matter how hard you press. The real tyrants are the denser woods with disturbed grain when the plane will not fail to be pulled into the wood, lifting shreds of unyielding fibres. But, sometimes the rewards for overcoming this problem are strikingly beautiful figuring. Within their trunks, trees rarely grow exactly vertically. Most spiral to some extent. In some, the spiral is so accentuated that in a small piece it is absolutely impossible to achieve a smooth surface with a plane, and you may be sure that the wood will react to "unwind" the spiral so that the piece will turn like a propeller. In some trees the spiral is clearly visible. Sweet chestnut with its furrowed bark often demonstrates this well. Tropical trees, especially African, tend to have a habit of spiraling regularly in alternating directions so that, if you saw a piece along the diameter, there is a striped effect. It is often very decorative, but it is very difficult to achieve a smooth surface with a plane.

Other secrets hidden within the wood are the aromas, mostly gorgeous but sometimes revolting. The many related and unrelated woods called cedar, as well as the original Cedar of Lebanon, all reveal a treasure trove of fragrance, although some people find this oppressive. Other woods smell of cinnamon or camphor; there is the rich resinous forest smell of pines and Douglas-fir, the sensuous smell of sandalwood, and the businesslike smell of teak. Beware of any sizeable piece of Viburnum wood with its khaki-brown heartwood – almost indefinitely it exhales the most revolting smell of cats' incontinence.

Planing a sample on the Victorian workbench in my workshop.

I have inherited a good collection of carpentry tools dating back to Victorian times and have supplemented these with other hand and power tools. While I use the power tools for other work, the samples are mostly prepared with hand tools except when power sanding is needed for the most recalcitrant woods. I find that the many acacias of the dry regions of Australia are the most difficult to finish. They are excessively hard and heavy (they sink unhesitatingly in water), slightly oily, and often have wildly distorted grain. Hand planing is impossible because the plane slides gracefully over the smooth grain or bites deeply into the disturbed grain. Power planers object whiningly or shatter blades. I have found the only way is to saw a little oversize, then power sand and scrape with a plane blade. The rewards are superb. The colours, luster, and cold marble-like textures are unmatched.

As the collection grew, I had to evolve a space-efficient yet easily accessible storage system. Another criterion to meet was that the samples were not exposed to daylight since ultraviolet rays change the colours, lightening some and darkening others, while generally removing the lustre or tarnishing them. Also, there must be good ventilation to prevent mould or fungal attack.

Our home is a former farmhouse with a small number of outbuildings including a row of former stables for the farm horses. This has become my sample storage building. I solved the display problem by creating double-sided stands of 30 tiers of shelves, each containing 30 samples laid flat. The shelves are just sufficiently high to store the samples and placed only a quarter of an inch above each other so as to exclude light but allow ventila-

The sample storage and display racks.

tion. This system has stood the test of time and growth. The racks are set out as if it were a library, and below and above the shelves are displayed the fascinating wealth of "fringe" items which 50 years of collecting have yielded. There are pine cones big and small (some as big as your head); pods two feet long, fruits of the kapok, baobab, and cocoa trees; and pieces of rare woods such as the frankincense tree and a piece of cypress wood from the middle of the Sahara Desert. Nature's quirks are represented by square, not round, cross sections of trees and others from trunks so fluted that the cross sections look like pieces of a jigsaw puzzle. Man's input is represented by such things as Nigerian, Balinese, and Maori carvings, a real aborigine boomerang, a Tudor sash pulley made completely of wood, huge Breton clogs (sabots), and a length of hollow drain pipe dug 150 years ago, out of the High Street in a town near Bristol.

Ancient and fossil wood has always interested me, and there are examples of petrified wood of several ages from Arizona, Libya, Australia, Argentina, and, nearer to home, from the Jurassic site at Portland, Dorset. At two extremes are amazing survivals of wood from Australia, not petrified but 10 million years old, and from an English coal mine beautifully preserved petrified leaves and flattened tree trunks dating back 250 million years.

Much younger (say only 5,000 to 8,000 years old) are pieces of bog oak and pine from the Fens of eastern England. When these trees collapsed into the soggy, airless peat, their wood was preserved and has turned up from time to time in astonishingly good condition. There is something special about the oak, however, because it is usually as black as ebony and, therefore, much sought after. It is an interesting chemical reaction, for oak is rich in tannin which reacts with iron in the peat and gives it the black appearance. Even in the short term, oak in contact with iron or steel and moisture will quickly turn to a blue/black ink colour.

Over the years I have had a steady stream of visitors, and many of them impromptu. Suppliers of samples from New Zealand, Australia, Argentina, Canada, America, South Africa, Holland, India, and Sweden have just dropped in to have a look.

The collection seems to intrigue the media. Our BBC local radio station ran a series of programmes about nature around us, called "Sounds Natural." The collection was featured on that and it led to some more general features about trees and wood and its uses locally, culminating in a programme called "Heart of Oak." This in turn led to taking part in two national radio programmes featuring what they surprisingly thought were peoples' slightly unusual habits. A similar programme, "Collectors Lot," on national television featured the collection in much the same way. A fascinating programme by the late Nick Darke featured things washed up on the north coast of Cornwall after drifting across the Atlantic, including lifebuoys from Canada and tropical seed pods from the West Indies. It was a pleasure to help with the identification of the wood in some of these intriguing things.

People listening to and watching these programmes and elsewhere were keeping me fed with a steady supply of new woods. In particular, I was helped tremendously by the owners of two estates in Cornwall, who not only provided me with a large number of suitable logs from their aging and ailing trees but also put me in touch with other helpful landowners. With their favoured gulf stream climate, Cornish gardens are famed for their collections of rhododendrons, magnolias, and camellias. Many of these, and a large number of other trees, were grown from seeds collected on expeditions to the Himalayas and China by the famous plant hunters at the turn of the 19th and 20th centuries. Sponsored by landowners since Cornwall's climate was so suitable, many growth trials were made there, but, of course, with the passage of time there was a steady dying-off. As well as other prized species, I was able to bring my species total up to 56 tree-rhododendrons, 25 magnolias, and 8 camellias.

To all of these helpers, I say many thanks.

Chapter 16
Great Woods of the World

Having shown my appreciation for so many who have helped, I am devoting this chapter to my appreciation of wood.

Wood is man's greatest and most useful, ever-renewable resource. It is eminently user-friendly and involves negligible pollution in its production and use. Wood has a thousand and one uses for strength, durability, hard-wearing endurance, stability, resonance, and resilience. Think of fenders, bridges, shovel and axe handles, violins and their bows, clarinets, wine barrels, ladders, truck bodies, screwdriver handles, surf boards, pallets, model airplanes, gates, piano keys, chairs, rafters, and mallets and you have still only scratched the surface of how wood can be used.

My interest in wood is as a decorative material. The range of beauty is infinite, for not only is every wood different, but every piece of every wood is different. The urge to drool over an exquisite piece of satinwood furniture is matched by the urge to touch it and run one's fingers over it, not only to feel the texture of it but to question how so apparently convoluted a surface can actually be quite flat.

There is nothing new in man's reverence for fine wood – it has lasted for thousands of years. It is arguable that the level of sophistication achieved by the ancient Egyptians was not exceeded until the great flowering of the classic era at the end of the 18th century and early part of the 19th century.

Volumes could be written, and have been written, about the world's woods, but the list which follows hones in on some of the finest – those with the highest drool factor. Some are well known universally; some are known only to the esoteric few. To keep it simple I have, on this occasion, generally not mentioned the botanic names.

Mahogany

The original, and some would say the only real, mahogany is the one that is found in the Greater Antilles and Bahamas of the West Indies. The wood we know as Cuban, Jamaican, or Spanish mahogany first attracted the attention of European discoverers in the mid-16th century. Cortez used it as a shipbuilding wood, but before long its beauty was appreciated; the oldest surviving use of it that I have heard is as a decorative wood in the Cathedral of Santo Domingo, dating back to 1514. By 1563, King Philip II of Spain was using it to adorn the Escorial Palace. The earliest known use in England was in Nottingham Castle in 1680, but it is almost certain that the wood was coming into the country long before then under the name of cedar. Cuban or West Indian mahogany was furniture wood par excellence. Suddenly, furniture makers who had been restrained by the size of walnut or oak trees had the freedom to use this wood of generous sizes, free of most defects, easy to work, and nearly sure to remain flat – and as if that was not enough, the rich red-brown colour and high figure gave beauty undreamed of before. There is no other wood that combines so many assets. Plentiful then, it is rare now, and largely irreplaceable.

The mahogany of the Upper Amazon, mainly Brazil, technically a true mahogany but a different species, combines many of the good qualities of the Cuban, but as Cuban is the aristocrat, Brazilian can only be regarded as middle class. African mahogany is also technically regarded as true mahogany, but its generally softer texture and "woolly" grain which resists the most conscientious efforts to achieve a good finish mean that it ought really to be deemed working class.

Human nature being what it is, many other red woods, some of them related to the real thing, some not at all, are called mahogany. Some are certainly nice woods and superior to the lower grades of "real" mahogany, but let the buyer beware!

Walnut

Before mahogany transformed the furniture making world, walnut had become the aristocratic wood. A most bountiful tree is the walnut with its fruit and with its wood. At home in south-east Europe and Asia Minor across to the Himalayas, the major supplies originate from plantations in Europe. Mature trees produce the best quality wood even though much of it can be quite bland. Experts argue about the best areas for the best walnut. The climate of southern Europe suits the tree best, and France and Italy are generally regarded as the finest sources; however, the climate of southern England seems to lend itself to eccentricities in walnut, a wood whose character is accentuated by defects. In mahogany, the figure depends upon the disposition of the grain. In walnut, it is the degree and disposition of the dark streaking which gives the wood its character. The crème-de-la-crème is produced by the root burrs or burls where the grain is wildly disturbed and the streaking patterns are at their best.

American black walnut is a true walnut but it is not a black wood; it is an attractive dark brown to purple colour. There other walnuts in the Americas spreading down to Argentina, but they are more of academic than decorative interest.

Naturally, such a prized wood also has its imitators. Queensland walnut, New Guinea walnut, and so-called Brazilian walnut (imbuia) all have a good resemblance to true walnut, even if not remotely related to it. African walnut is neither related nor really very similar.

Oak

Oak has so many uses that decoration is just one way in which it excels. Before mahogany, they used walnut, and before walnut, oak. Perhaps nowhere more than in England is this appreciated. That nice, chunky Jacobean and Tudor furniture is much valued now. The great attraction of oak is in producing the timber so that the large rays are exposed to the best advantage. These rays, which are highly functional parts for the tree, are extremely decorative for the timber. They radiate out from the centre of the tree, and they are seen at their best if the wood is sawn radially.

Although we English tend to think of the oak as being one archetypal tree, there are actually two species in Britain which produce indistinguishable timber. They are found over much of Europe where several more species occur. Many more occur in America, too. In fact, the oak's real centre is further south than Britain. There are no less than 400 different oaks with a large concentration in Mexico and eastern Asia, including the warm dry regions. The best woods, however, come from the north temperate regions, with American white and red oaks being highly prized.

Some other trees, especially in the Southern hemisphere, have woods with the broad rays, but none more than in Australia. In fact, in Australia, one group, the silky oaks, produces superb wood with more numerous but smaller rays than "true oak," and another group, the she-oaks, has wood which superficially resemble the woods of the evergreen true oaks.

Rosewoods

Related to rose trees only because the wood has the fragrance of roses when worked, rosewoods are actually trees of the bean or legume family. Essentially tropical, there are very many species of the genus Dalbergia, only a small proportion of which produce the fine timber.

Most would say that the finest is the Rio rosewood of Brazil. This is the wood with those black streaks which, at their best, wander illogically to produce entrancing patterns. Indian rosewood is a rich purple with less streaking, while Honduras rosewood is harder with a rich lustre. Incidentally, it is the favourite wood for marimbas because of its special resonance.

Other Dalbergias produce the spectacular Brazilian tulipwood which has a smooth cream-coloured texture striped with bright red streaks, while kingwood, also from Brazil, is aptly named for its outstanding appearance of dense, rich purple with streaks of black. Cocobolo is excessively hard with an amazing metallic lustre and a cavalcade of reds, browns, and blacks. African blackwood outdoes ebony in its pure blackness and cold smooth finish, and it is the favourite wood for bagpipes.

Ebonies

I find that most people think there is only one ebony and that it is pure black. Some are even surprised that it is actually wood. In fact, there are several different ebonies and by no means are all of them pure black. All of the true ebonies are produced by species of the Ebenaceae family and nearly all by species of the genus Diospyros within it. But there are 400 species of Diospyros and perhaps only 50 produce wood that is at all black. Commercial supplies of these pure black woods come from India, Sri Lanka, and West Africa, but others occur in Mauritius, Madagascar, Thailand, and Vietnam. The decorative uses are legion, and the story of its use extends to the boundaries of history.

Many connoisseurs prefer the streaked ebonies produced by a few other Diospyros species, such as the regular striped Macassar ebony of Indonesia and the tigerwood of Malaysia or the wildly irregular streaks of Andaman marblewood, Philippine ebony, and New Guinea ebony. Most highly regarded is the beautifully streaked and mottled calamander or coromandel wood of India and Sri Lanka which is now virtually unobtainable.

A few more yield occasional black areas, such as the persimmon of America, and the vast majority yield none at all.

Satinwood

Real satinwood comes from southern India and Sri Lanka. It is a fine, rich creamy-yellow wood which at its best is so figured that it resembles the finest satin. Once seen it is never forgotten, but the wood is rarely seen now.

Teak

There is only one true teak and no really successful pretender to the title. It occurs naturally in India, Burma, and Thailand in monsoon climates. Like oak, of which it might be said to be a tropical equivalent (although not remotely related and not of similar appearance), teak has a host of uses. In Victorian times, teak was seen much more as a shipbuilding timber. Now teak is rightly seen as one of the world's prime decorative woods. The rich brown self-lubricating appearance is unmistakable. Supplies from natural sources are now effectively limited to Burma, but so valuable is the wood that any suitable tropical country grows it, such as Indonesia, Guyana, Trinidad, Nigeria, and Sierra Leone.

Boxwood

Box trees occur from southern England in the hills, through to the Himalayas and Japan. Despite their small size, the wood is much prized for its outstandingly hard, smooth yellow appearance. It is a favourite wood for carving.

~ ~ ~ ~ ~ ~ ~ ~ ~ ~ ~ ~

Briefer mention can be made of the woods which follow – these are either less well known or hardly known at all, but they include just a few of the other fine woods of the world, which craftsmen may like to try.

Berberis and Mahonia

Perhaps gardeners will be surprised to see these included in the list as both are from shrubs or small trees. Related to each other, no other tree has such a strident yellow wood, and it can be made into striking curios.

Blackthorn

Perhaps another surprise is this rampant, usually small, European bush which is so often regarded as a weed. But, if left alone for a long time, blackthorn, which bears sloe berries, will grow to a height of 30 feet and have a small heartwood richly coloured with red, dark green, and purple.

Bocote and Ziricote

These two related Central American woods are quite similar in appearance. Although not related to Rio rosewood, they could almost pass for it with their firm texture, natural sheen, and nice dark streaking.

Brazilwood

Although really a wood best used for dyeing, brazilwood is attractive in being a bright flamboyant orange. The country of Brazil is named after it, not vice versa.

Coconut palm or porcupine wood

Palm wood is interesting because it has a totally different structure than other woods. Scientists say these are vascular bundles in background tissue. Hardness varies from very soft to quite hard and wood lovers single out the hardest woods, like the coconut palm, because the vascular bundle effect is like porcupine quills. The decorative effect is unmistakable.

Desert ironwood

This small tree, which flourishes in the dry regions of southern Arizona and northern Mexico, produces a marvelously hard and heavy dark brown wood. It is the favourite carving wood of northern Mexican Indians, although the supply now is very limited.

Lacewood

The London plane tree wood, and its equivalent the sycamore in America, produces a very striking ray pattern resembling lace when sawn radially. Cut any other way there is no such effect.

European olive

Like walnut, olive is a bountiful tree. When its fruit-bearing days are over (up to 1,000 years), the ancient olive tree can be harvested for its superb, fragrant wood. Cream coloured with a wildly unplanned array of brown and black streaks, it is a favourite turning wood.

Olive ash

The European ash is a close relative of the olive. Old trees produce a core heartwood which not only looks very much like olive but smells like it too.

Oysterwood

This wood of a small tree of Florida and the West Indies has some resemblance to olive, but it is generally darker and with a higher lustre.

Padauk

The original padauk is a Burmese wood, but similar woods are found in the Andaman Islands and Africa. No other wood is like it, for the most highly coloured is bright red, although some prefer the more restrained colours of its Amboyna species.

Purpleheart

Several species of Peltogyne of tropical America produce this purple wood. At its very best, the purple is so strident as to be almost unreal.

Thuya

Roman Emperors paid vast sums to have furniture made of the burrs or burls of this dry-zone North African coniferous tree. The figured burrs, which mostly grow underground, are still highly valued.

Yew

Yew trees reach a great age and their shape is often quite distorted, but this gives their wood a great deal of character. The rich brown heartwood contrasts with the white sapwood and the artistic shapes produced from the fluted trunks are very ornamental.

and finally –

Snakewood

Also known as letterwood, snakewood is my favourite wood. There is nothing remotely resembling it, and only the best is good enough. At its best, the heartwood is a bright rich brown, very dense, smooth and with a glossy lustre. Set into this background are patterns of prominent black blotches that look like letters or give the overall effect of a snake's skin. In its native Guyana, the tree is not very common; only a small heartwood is produced and only some of the heartwood is blotched, and only some of that is blotched consistently enough to give the snakeskin effect. Not surprisingly, this is probably the most expensive wood in the world.

Chapter 17
Across the Pond

It seems very long ago in this narrative that I was making my European expeditions but that only mirrors the facts. Fifteen years had passed since my trip to Crete. The family had grown up and the wood collection had been growing steadily without expeditions.

By 1991, however, I had been musing about a perfect pipe-dream holiday that would combine seeing some of the most interesting trees, collecting a few samples, and seeing some of the most spectacular scenery in the world. For this pipe-dream, I had narrowed the options down to six when it became clear that the southwest of the United States was emerging as the clear winner.

For several years I had been corresponding with International Wood Collectors Society member Alan Curtis in Oregon, one of the most dedicated collectors of wood specimens. I wrote to him asking for a little guidance on the area and asked him if I could drop in to see him after the trip when perhaps he could help me and the returning aircraft by sawing down some of the trophies to more manageable sizes. His reply was very helpful and enthusiastic. Alan suggested that he and his wife Mary Ann drive their camper or RV (recreational vehicle) down to Phoenix, Arizona and meet me there. So in July 1991 that is exactly what happened.

We met at the airport at Phoenix, Arizona. Known as Pumkinville in the 1870s, Phoenix rose from the ashes of the old Hohokam Indian settlement in the desert to become today's bustling, prosperous, and well-watered city of around 2 million people. On a map and from the air, it looks like some gigantic crossword puzzle, but so generously is it spread that gridlock must surely be a long time away.

The first night we stayed in a motel but from then on we were to camp at a succession of well-chosen places. Their camper, built for two, was really too small to accommodate me, so for me they had a little tent which I really enjoyed.

On our first day as we left the motel in Phoenix with the sun turning up its blistering heat, we drove off west and south to reach the Organ Pipe Cactus National Monument on the Mexican border.

The cactus country of the Sonoran Desert in southern Arizona extends down into the Mexican State of Sonora. One would expect the country, which is bruised by relentless sun and rationed to nine or ten inches of erratic rain in a year, to be a lifeless waste or at most covered by the occasional miserable plant. But it is not like that. Dominating the scene are the huge Saguaro cacti so familiar in border Westerns. Their single fluted trunks armed with prickles soar up to 30, 40, or even 50 feet with a few arms sticking out like some deranged robot. Right at the tips of the trunk and branches are a mass of waxy white flowers which are followed by juicy fruits. Many are disfigured but not harmed by holes driven in their sides by the Gila woodpecker which makes its home there. At this southern point are also the organ pipe cacti. These are quite different in appearance with a cluster of several stems rising to 20 feet and looking like a church organ. Then there are the barrel cacti, more like a large prickly ball resting on the floor of the desert, and the jumping cholla cacti which are akin to prickly pears. Excessively prickly, they dangle chains of barbed fruit which any passing animal (including human) is likely to carry away to a safe spot for germination. To complete the scene, there are ocotillos, which are not cacti but look like clumps of prickly walking sticks topped with spectacular red flowers, and finally the ubiquitous creosote bushes. Toughest of all, these clusters of stems rarely more than five feet tall have tiny leaves which reek of creosote when touched by hand or by rain. Underground the spreading root system penetrates the ground with toxic substances to inhibit any competitor. So the ground is covered for mile upon mile of neatly spaced creosote bushes.

When the rains come, they come in torrents, and the gravelly ground is unable to absorb the deluge which takes the lowest points, the dry river beds or washes, and produces a tidal wave, the flash flood. Alongside the washes are the other trees of the desert. Many are of the bean family like the mesquite, paloverde, and desert ironwood.

There were no flash floods while we were there, and the temperature climbed to a merciless, parching 105°F which is by no means unusual. We moved across the Papago Indian Reservation to Tucson, Arizona and then on into Mexico for a day's outing at Nogales. Back in Arizona, we trekked east into Tombstone, an absolute must for Wild West fans. Here at the OK Corral, the Earps and Doc Holliday slugged it out with the Clantons, McLaurys, and Sheriff Behan, although this famous fight was really only a

Beside a cholla cactus on the Arizona/Mexico border, 1991.

A 50-foot-tall Saguaro cactus in southern Arizona.

Organ-pipe and Saguaro cacti with creosote bushes in southern Arizona.

battle in a long-running war which claimed many more lives. An OK Corral, original or reproduction, still exists at Tombstone and in it are lifesize figures of the fighters, each duly labelled. One does not dwell too long on the fact that the gunfight actually took place on the street outside.

Arizona is brim-full of natural and man-made curiosities. At Morenci is a vast copper mine – a huge quarry of an open-cast mine where a copper-laden mountainside is being shaved away progressively to oblivion. Near Holbrook is the Petrified Forest National Park, the largest concentration of petrified wood in the world. On a 16-mile drive you can see huge areas of wood turned to stone. This wood was part of trees growing 200 million years ago and now is transformed into garish stone colours by liquid silica under pressure. Complete tree trunks 150 feet long are exposed to see. Further on is the Painted Desert, Canyon de Chelly of the ancient Indian houses sheltered under precipitous cliffs, Monument Valley of "Stagecoach" fame, and Meteor Crater where 50,000 years ago a meteor crashed into the plain, punching a hole 500 miles deep and 1 mile in diameter, displacing 300 million tons of rock.

But, of course, the Grand Canyon is Arizona's greatest attraction. We stayed there for two days marvelling at the sights. It is difficult to take in that it is 200 miles long, 13 miles wide and 1 mile deep. Considering it is such a tourist attraction and has been for many years, man has been remarkably unobtrusive. Only hairline hairpin trails lead down to the Colorado River seething below and back up to the other rim. We hiked for two miles into the chasm, and struggled back again.

North of the Grand Canyon into Utah, the red country begins and for me one of the finest sights in the world is Bryce Canyon. It is not actually a canyon but a huge cliff face which is eroding progressively to reveal countless rows of red columns pierced by chasms. The result is spectacular.

Bryce Canyon, Utah.

Death Valley lived up to its baking-hot reputation even if it was a mere 110°F that day. We camped at its north end to find that the searing wind blew up the valley in the day, stopped for a rest in the late afternoon, and then blustered down the valley, parching our lips and making it very difficult to erect my tent.

A bristlecone pine high in the White Mountains, California. One of these trees is the oldest living thing in the world – 4,600 years old and still alive.

Perched on the top of the White Mountains on the California/Nevada border are the oldest living things in the world. At between 10,000 and 11,000 feet, in desperately poor soil and with little rainfall, grow the bristlecone pines. Very little else grows because the regime is so severe and bristlecones dominate the mountainsides. Their growth is infinitesimally slow, perhaps an inch in height a year, but it is steady and resilient. The climate cannot support complete ancient trees so perhaps only 10 percent of the tree is still alive, but that part is very much alive. In amongst the groves of the trees is Methuselah, the oldest bristlecone pine, which is known to be 4,600 years old. How is it known? It is not by carbon dating but by drilling into the trunk with an increment borer, taking out a minute core, and laboriously counting each growth ring. So harsh are the conditions that fungi do not thrive, and the wood remains sound for centuries. Only the wind whipping up sand and ice-particles abrades the trunks, polishing them into rich honey colours. In gullies, centuries of cones are piled in their thousands, and dead wood has been picked up, measured with an increment borer, and found to be 8,000 years old and still intact. The core boring is not just a means of counting years but also a record of climate; each sequence tells a story and the old dead wood sequences can be tagged onto those of living trees. So accurate a time measurement is it

General Sherman, the biggest tree in the world, in Sequoia National Park, California.

that, in the early years of carbon dating, it was possible to compare those readings with ring counts and establish that the carbon dates needed correcting.

We took our camper to the highest attainable point at 11,000 feet, almost at the limit of the bristlecones. It came through the test with honour, only protesting with overheating brakes on the way down again.

California not only possesses the oldest living thing in the world but also the tallest – a Californian redwood – and the biggest – the General Sherman Tree, a giant sequoia and a relative of the coast redwoods. It is one of the group of these trees on the western side of the Sierra Nevada. It is only a stone's throw from the White Mountains to the Sierra Nevada, but such a barrier do these saw-toothed mountains present that we had a long, laborious, and hot detour around the south to be able to climb up to the Sequoia and King's Canyon National Parks to find General Sherman, General Grant, and their friends. Here amid cinnamon-barked trees towering in clusters above is one of the world's most humbling experiences – to stand at the base of the General Sherman knowing that it is 270 feet tall and 100 feet around its base. Humbling, and sobering too, to compare the destruction we can cause in our three score years and ten, with the vulnerability of such a giant of 2,700 years.

Thus humbled we headed off down the mountain to San Francisco. It was to be the grand finale of all the expeditions.

Chapter 18
Encore

But it did not work out like that. A year before that trip-of-a-lifetime, I had written to the Forestry Research Institute in Leningrad (now St. Petersburg, Russia) asking if they could help me with samples of a few more Soviet Union timbers. As before, they had been most helpful and sent me several, but they said they did not have the woods from the Caucasus region which I had requested, and suggested I should write to a professor in Tbilisi, Georgia. This I did without delay, but it was not until 18 months later, in February 1992, that I received a letter from this good man. "I wonder," he wrote, "if you have received the samples of Caucasus woods that I had sent you."

I replied with regret I had not and would it be possible to send me some more as I really very much needed them. Airmail correspondence from Tbilisi was quicker than surface but, at up to two months at that time, it constrained correspondence somewhat. Eventually in May I had a reply saying how sorry he was that I had not received the samples. They had been having a bad time, he said, and parcel post was so uncertain that it would not be wise to send replacement samples since they would probably not arrive. "I hope." he said, "the situation will stabilise soon; we'll reach good inter-relations between our countries and I'll have the opportunity to see you here, in Tbilisi." Posthaste I wrote back and said in effect, "How about now?" The post was unusually hasty and in only two months I had a marvellous letter from the professor, whom I shall call Giorgio, saying this:

> "I received your cordial letter, thank you very much. My family and I would be very glad to have you in Tbilisi in September. In present, because of the destabilisation of the situation in Georgia, everything is more complicated, but that will not prevent us to make you 'feel our hospitality' (as he phrased it). We shall show you Tbilisi and our mountain forests. I would like you to live in my house in Tbilisi and in my country house in a mountain resort."

Perhaps it was fortunate that I did not really know much about "the destabilisation of the situation in Georgia" because, if I had, I might not have ventured as I did. But if I had not ventured, I would not have had a unique experience.

In the early throes of the break-up of the Soviet Union, Georgia had gained independence, at least in theory, from Russia. Even so, a President had been imposed on them who they thought was little more than a puppet of Russia. In this volatile country, the Georgians had rebelled and sent the army into the middle of Tbilisi to Rustaveli Avenue, the centre of government. In a relentless barrage, tanks and guns came up the avenue and smashed to pieces the military headquarters and severely damaged many other administrative buildings, especially in Lenin Square (now Freedom Square), and even the elegant Hotel Tbilisi. This had resulted in the overthrow of the new regime, to Georgia's delight, but also absolute chaos with instability, electricity and water failures, and crippling shortages of petrol and food, bringing Georgia almost to its knees. The revolution had happened a few months before my visit and the resultant chaos was at its worst when I, in blissful ignorance, decided to visit.

Even the nerve-racking, grinding bureaucracy of such a visit and the horrendous communication problems did not deter me. I knew a good travel agent specialising in Soviet holidays and together we sliced our way through official invitations, visa, tickets, and transfers. Phones were busy between Cornwall, Manchester, Moscow, and Tbilisi, and one of the very few fax machines in Tbilisi was put to special use for the occasion.

On September 6, with just two weeks to go, I had not been able to get any plane tickets, visa, or insurance. From London to Moscow by British Airways was not a problem, but from Moscow to Tbilisi remained uncertain. They had made a provisional booking for me but it couldn't be anything more than provisional as Aeroflot was short of fuel and spare parts. At that moment, the agents had 50 tourists stranded in Moscow because their flight to Poland had been cancelled due to lack of fuel and there was no overnight accommodation in Moscow.

By September 20, I had the visa, the insurance, and my tickets from London to Moscow and return. In addition to that, a car had been arranged to take me from the International Airport to the Internal Flights Airport in Moscow. My flight from Moscow to Tbilisi had been confirmed, but this was after my planned flight was cancelled and I had to go on a different one which would make my transfer in Moscow very tight. For this purpose the agent gave me a document of many pages with tips on how to short-cut the bureaucracy (including "go to the front of any queues"). I assumed all would be okay, so I rang Giorgio to tell him my proposed arrival time. This was somewhat reassuring, even if he seemed to have great difficulty in understanding what I was saying in my phonetic Russian from my phrasebook.

I had been trying not to be anxious about the Moscow transfer, but I had just read a feature about holidaying in Russia which sounded moderately reassuring until the piece about security – "In the mountains of Svanetia (in the Georgian Caucasus) this summer, entire trekking parties have been held up at gunpoint and relieved of their valuables and a number of Russian tourists were killed in crossfire two weeks ago at Sukhumi. Extreme caution should be observed by visitors to the southern and eastern Caucasus (Georgia, Armenia, Azerbaijan, and Chechnia)."

~ ~ ~ ~ ~ ~ ~ ~ ~ ~ ~ ~

Midnight Plane to Georgia

On the next day, September 21, I was safely on the British Airways flight to Moscow, but it took off, rather worringly, 45 minutes late. It seemed like no time before we began our descent to Sheremetyevo Airport, which looked quite respectable. I had been warned to expect difficult immigration and customs, but it was all very quick and easy, even if the staff had a rather belligerent manner. Soon I was looking for the car driver who was to take me to Vnukovo, the Moscow airport for internal flights, and who should have my flight tickets to Tbilisi. Luckily I soon found him and he had the ticket. He spoke hardly a word of English, which just about equaled my Russian, but he was friendly and got me to his car, a decrepit Moskvitch. In this he drove me to Vnukovo along roads that were obviously "main" but not dual – just very wide with no lanes marked at all. All of the trucks and cars were predictably worn-out as were most of the buildings. The new buildings were dreary multi-storey.

Eventually, after skirting Moscow through rather pleasant birch and pine forests, we reached Vnukovo Airport. It was depressingly dowdy. It must have looked good once, but now it had not had a coat of paint for years. It had the air of a run-down, provincial airport in Outer Mongolia. In what laughingly passed for a departure lounge, only a tiny proportion of the neon strip lights were working, so it was getting progressively darker.

I kept a diary of the trip since I thought there might be a few interesting events, and what follows are a few of the items I recorded.

The plane was delayed – more than once! Nobody spoke a word of English; the departure board headings were all in English, but this was of little use to anyone because no flights were listed. Everybody seemed totally used to this though, and quite resigned to it all.

About 15 minutes before the delayed take-off time, we were summoned to the flight. The only other passengers at that point were, believe it or not, two British people – jugglers. They attended the World Juggling Jamboree in Tbilisi the previous year, staying with locals, and they liked it so much they were coming back.

In mighty contrast to London Heathrow security, we strolled across the tarmac to the plane, keeping a wary eye on manoeuvring jets, and climbed up the rickety steps to the plane. The plane was unbelievably squalid inside, with a nasty, rancid smell and the look of the 1950s plane that it probably was. The floors were clad with loose fragments of oily carpet of three different colours and the seats were ancient.

By 25 minutes after the new take-off time, there was still no sign of activity. What I has hoping was the crew kept sauntering on and off the plane or lounging at the front out of sheer boredom. As we, the only three passengers on this 150-seat plane, continued to wait, I could not help but wonder what happens now – do we wait here indefinitely until the numbers are made up?

Finally a flurry of passengers, all Russian or Georgian, were allowed onto the plane but not allowed to sit with us despite their fervent protests! One man, however, slunk past the hostess to sit with us and others soon followed like a horde, spilling into our part and bringing impossible loads, filling up the plane with their strange packages, strange guttural utterances, and strange aromas. With no booked seats, they piled in everywhere – but they were polite and friendly.

After more than an hour of waiting with the engines running, we finally took off with a full load of happy, outgoing people. My "neighbours" on the flight were teachers in Tbilisi who spoke a little English and were very friendly. I learned that they earned 1,500 Roubles per month (about £4), in contrast to a policeman who earns big money at 6,000 Roubles (about £15) a month.

The four air hostesses on the flight were very well presented, but they did virtually no work at all, choosing instead to sit in their section of the plane, chatting and smoking. Eventually one brought some mineral water in small brown plastic cups – the only thing offered on the flight of 2-1/2 hours.

On the flight I also learned of the strange double standards in pricing. I was paying 68,750 Roubles (US $275) for my return flight, whereas the Russians were paying only 3,700 Roubles – I was worth 18 Russians! No wonder the tourists were allowed on first and given the best seats such as they were. This was a new meaning for first class.

Eventually, in the pitch dark at midnight, we landed at Tbilisi, although I could not see a building with a sign confirming this. We all ambled casually across the tarmac and out of a small doorway in the fence, to face the throng of friends.

But then I noticed that I had, in fact, left the airport, lost my English juggling friends, and had nowhere to claim my luggage. So I wandered back in again where a man, casually holding a machine gun, showed me to a baggage hall which I should have entered. Inside was a carousel with baggage, but not mine, and now my English friends reappeared. Then someone said aloud that this baggage was not from the Moscow flight. "No, it's not this carousel; it's that pile of baggage over by the Intourist Hall on a truck." And, lo and behold, my baggage and Giorgio, my host, with his brother, were finally before me. Hooray, mission accomplished.

At Giorgio's car, an elderly Lada no different from many others except for having a smashed windscreen, we piled in, bag and baggage, and set off on the 25 kilometres to Tbilisi. Giorgio drove in a quite a professorial way, chatting energetically to his brother, and we made our first faltering attempts to reconcile our language gap.

We had not driven far before we were halted by a blockade formed of burnt-out vehicles manned by a motley crew of gun-wielding desperadoes. At this time in the night, it was certainly not good news. They searched the car, apparently looking for guns and ammunition. Giorgio informed them that he was bringing an English professor from the airport. We were permitted to pass by without hindrances, but I had no idea what were the credentials of these people. Were they the face of authority or were they anarchist terrorists? I had no means of finding out, but we passed through three more blockades before reaching Tbilisi.

The city outskirts looked ramshackle, but they improved steadily. We soon reached imposing, tree-lined streets and then into Freedom (formerly Lenin) Square, then up Rustaveli Avenue. Here I got a nasty shock. The level of damage to major buildings from the three-week war of Tbilisi was absolutely horrendous. There was a huge area of blackened shells of fine buildings – quite like the Blitz. It was too dark to see much of it then, but I was sure I would see it, and more, in the daylight.

In contrast, Giorgio's street was in a nice area of Tbilisi and he proudly showed me the entrance to his block. Among the three-storey apartments, we selected his building and climbed up the shallow flight of stairs. There were broken windows, but once into his apartment it was magnificent. There were very big bedrooms and a large dining room with robust, Yugoslavian furniture. Giorgio and his wife, a teacher at the Mathematics Institute, gave me a wonderful greeting and I brought them presents of whiskey, cigarettes, and perfume. While I thought we would go to bed next, no. Even if it was 1:00 am, we had our dinner with bread, meat balls, spiced aubergines, and cheese. After formal toasts of wine and chacha, the Georgian fire-water, we finally went to bed at 3:00 am.

In the morning after breakfast, Giorgio planned to take me on a tour of Tbilisi, but he received a phone call telling him where he could refuel his car by queueing up at a tanker which had stopped. No fuel ever reached a station as it was sold straight from the tanker by the roadside. I began to understand the poverty that had struck Tbilisi. Two years before, the economy was stable and people had enough to live on. Now there was abject poverty. Before us was an open museum of traditional old houses of all regions, obviously set out for tourists. But now I was the only one, and the houses were derelict and abandoned. While I was taking photographs, a man said to Giorgio that he should not let me see these things. Georgians are proud.

Republic of Georgia. One of the many buildings almost destroyed in the Siege of Tbilisi, December 1991.

As we continued about town, I could really see the poverty. There were several closed shops; those that were open had a handful of food essentials for sale but otherwise nothing. There were many people ambling around aimlessly, however. We visited a market with meat and trussed chickens for sale at exorbitant rates. One chicken was one week's pay. I learned that a pair of shoes cost one month's pay. How could they afford them? They couldn't; they all wore old ones.

I asked to see Rustaveli Avenue where the fighting was, and I was deeply shocked. The damage was really bad. It was not just the Parliament Buildings, but buildings over a large area were totally destroyed or burnt out. The buildings still standing were pock-marked with thousands of shell and bullet holes. It would take years for this poor country to repair the damage.

It was clear, however, from the imposing buildings that were not damaged and the majestic streets in its centre that Tbilisi had enjoyed prosperity and a high standard of culture. Perhaps in return for this they tolerated the ardours of a foreign, Russian autocratic regime. Giorgio had obviously enjoyed prosperous living standards. But Georgian freedom had been bought at a great price and the loss of living standards was a bitter pill.

The evening brought some excitement that was unusual to me. I was sitting writing in my diary when suddenly I heard shots in the street below – about 50. It was very frightening. Also I heard screams and saw several police cars and vans. Several people came out onto their balconies and there was a cluster of people in the street looking at something just out of sight around the corner. Was it a robbery, armed uprising, a shoot out? Who knows? Anybody killed – who knows. It's anarchy, and things were not like this under the Communist Regime, I was told. Perhaps we would know more in the morning or perhaps not – that was life now.

On the following day, September 23, Giorgio announced:

> "Today, Richard, today, we are going to a cement works."
>
> "Oh, how nice!" I replied.
>
> "Yes, my professor colleagues and I are going to give some advice on avoiding erosion and pollution. We shall also take experts on rivers."

We set off early, and one of Giorgio's colleagues was already standing by his car as we reached it. We collected two more in other districts of Tbilisi. This also provided me with an opportunity to see more. I must say that, away from the graceful city centre, the suburbs were very squalid. There was rubbish everywhere, roads were deeply pitted, and buildings were grimy. There were feral dogs everywhere, and at night they hunted in packs, marauding and barking in the streets. There were road works which had been started and then abandoned. The attitude toward traffic lights was condescending, and Giorgio remarked that some red lights are redder than others!

Soon we were heading north out of Tbilisi on the "Georgian Military Highway." We were stopped by a blockade but allowed to go through without comment while others had their car contents checked. Following a riverside, we steadily climbed up into the mountains, which were very bare and unimposing, a scene not helped by heavy, overcast weather. We met another car with the other experts, and our group then totaled seven. Maps came out and then a cacophony began with all of them talking at once at a furious rate. After two hours, they seemed satisfied. We moved on to the cement works being built (or was it falling down?), a very grim sight in the middle of nowhere but with a charming young man as manager. Then we joined a main road and ended up in another village. We drove down to a spring and a river where they checked the water for purity prior to the cement works beginning production. Then finally it was time for lunch.

For lunch, we sat at a primitive steel table and out came the goodies – round bread, melons, tomatoes, cheese, butter, and of course chacha, the Georgian firewater of unknown strength. They apologised for the poor food, but I protested that it was wonderful. Then the tamada (toastmaster) proposed a toast to Georgia and their distinguished guest. I proposed one for peace and prosperity in Georgia – and so on.

"You must finish the glass each time" they said.

The chacha drinking session which had an unusual ending, in the Republic of Georgia, 1992.

On one occasion, you must finish two glasses. Unfortunately, it was soon clear that I could not stomach the chacha, which seemed to be very high proof. I felt violently sick, as though I was dying. They also worried that I was, and perhaps I was a having a heart attack. Very concerned, they rushed me into a doctor's surgery, where I was laid out on a table, given an injection in my bottom, my heart massaged, and my blood pressure checked. There was no such problem, of course, but I was desperately sick over and over again. Eventually though, I was fit enough to be driven home, when I slept it off for ten hours and woke up with virtually no headache.

More shooting occurred that evening, but measured shots. Giorgio determined that it was the police, and he seemed relatively unconcerned. Perhaps they were just warning shots or blanks.

Fortunately, the next day I had fully recovered. Because of the electricity shortage, our plan was to leave Tbilisi and visit Giorgio's house in the mountains about 150 miles away. It was pouring with rain and the Soviet-style features were not a pretty sight, but we drove off in this rain and at length began to climb up from the 1,500 foot plain we had crossed to the southern mountains. Instantly the scenery improved, except in the villages where the grim factories and multi-storey apartment blocks scarred the countryside. By the time we reached the town of Borjomi, it was dark, but the locals did not seem to want to use headlights (or lights at all) until they absolutely had to do so. This added a lot to the hazards of driving in Georgia.

We climbed further up and the weather continued to deteriorate as we entered the clouds. We were in mixed woodland of Caucasian spruce and various hardwoods dominated by beech and oak.

The road degenerated into an unmade surface with horrendous potholes. There were long periods of forests, then villages where we turned left and right, until at last they pronounced that we had arrived. It was actually a lovely little village, and Giorgio's house was impressive but spartanly basic. Inside, the house was on two floors with an external staircase. There were two large downstairs rooms, the first of which was a kitchen/dining room/sitting room with a steel wood burner, a long table, and two benches. There was a propane gas stove, but it didn't work. We settled in and soon we had a group of five men and one woman enjoying the usual fare. We ate and drank until 11:00 and then I moved into the other downstairs room, a large bedroom with three beds, and soon I was asleep.

It rained all night and was still raining in the morning. There was a lot of low cloud cover but we could see snow on the higher mountains – the first of the winter. The village scene was like something out of a Western picture. The houses were built of planks of spruce and pine, and there was a primitive privy in the garden. Horses and carts were plodding through lakes of water filling the potholes, and all kinds of animals were running wild and had to fend for themselves – horses, cows, pigs, dogs, cats. It was like a time-warp of 200 years to see all of this, but the friendship was wonderful and the scenery spectacular. We were 6,000 feet up, and here it was flat, but all around us were greater heights and especially to the south they rose very high and were deeply clad with snow.

After breakfast we set off in the drizzle, in a decrepit Moskvitch which really was held together by pieces of string. It wouldn't start at first, but eventually the gallant little engine sprung (well, groaned) into life and we baled out the water which had trickled in through the roof. Off we went along the appalling lake-roads through the village. We went back through splendid spruce and pine forests mixed with eastern beech, Caucasian maple, and hornbeam, and quite frequent and large Caucasian pear and apple trees.

On the way to the town of Borjomi, we stopped at three Institutes where I was introduced as an English professor and, as a result, given a marvellous reception. All had excellent museums which they proudly showed me.

We headed back to the village. At last it had stopped raining and the clouds were clearing, but there was a lot more snow on the mountains.

There was a nice meal waiting for us at our next-door neighbour's house. There were numerous toasts of course, and my stomach was becoming accustomed to the chacha. The television was on and it seemed to be all about fighting in Abkhazia (in western Georgia, which was a no-go area).

When I awoke the following day, I saw almost clear skies and the sun shining down on the most marvellous mountain scenery. All around the village was a large flat area of fields which belonged to the collective farm. It was studded with occasional pines and wild pears. Beyond the plain were steep, wooded slopes, and above those was the blanket of snow to peak at about 9,000 feet. It was a glorious scene and very dramatic.

Gradually the hunters gathered for an outing, checking their guns and cartridge belts. We had the usual breakfast laced with wine, brandy, and chacha, and then set off on our expedition. Walking across the flat plain, past the cows, geese, horses, and pigs, we were a party of seven – six with guns and one with a saw – to hunt for things lurking in the forest. We were accompanied by three hunting dogs.

With my host, Giorgio, high in the mountains of the Republic of Georgia looking for things lurking in the forest.

At the end of the plain, we crossed a river and began the climb. The numbers of species and richness of vegetation were at their best here. It was fairly moist country, so there were frequent alders and willows. The dominant species, however, was Eastern beech. A large tree and with its larger leaves and better branching, the Eastern beech is arguably a nicer tree than our English beech. Mixed with this were Caucasian species of elm, pear, apple, maple, oak, and birch as well as aspens and rowans.

As we climbed, we entered an area of virtually pure beech. The forest was fairly open and populated by trees of all ages, from old and very decayed to lush young growth. It was steep, but gradually we climbed up out of the beech and into maple, birch, and rowan. Then there were good views down to the village and up to the snow-covered mountains around us, making for a truly grand sight.

Upwards we went until the forest cover became thinner and thinner, eventually restricted to the birch and rowan. There were some nice sub-alpine flowers and meadows. Only three of us continued to climb up to the snow-line and there, exactly at the snow-line, was a carpet about three to four feet tall of naturally growing *Rhododendron ponticum* – the species that runs riot when unsuitably planted in England.

Moving back down again, we found a spot to make a half-hearted fire and stack the shotguns. Out came the inevitable food – bread, sardines in a tin, cheese, tomatoes, and of course chacha. This time, however, the complicated, voluminous, and garrulous formal toasts were drunk from a single cow horn "glass" which was passed around. Each person must drain it and hold it up to his ear to prove it. The tamada (toastmaster) was positively manic. He became more and more worked up in a frenzied diatribe at the top of his voice and allowed not a word of interruption. I managed to cope with limited amounts of chacha and proposed toasts of my own.

Finally after say two hours of fun (of which I could not understand a word), we broke up the picnic and headed back to the village by another route, making a circle. Fairly quickly we descended into the beech woods. After that we entered a pine wood and, in amongst the pines, I found the Caucasian pear and apple that I had been seeking. Now I realised just how kind these Georgians could be. One of our party, Vanod, was the village carpenter, so he did all of the harvesting for me, meeting my minutest needs. He was obliging to the point of embarrassment. He even took the samples home to machine-plane them for me. I was overwhelmed by all this kindness and didn't quite know what to do. Every time I tried to offer to pay or buy food for the household, they became mildly offended and stated that it is a Georgian tradition that the guest never pays.

No sooner were we back, then off we went to Waliko's house (Vanod's son) for another binge. The food was the best yet – aubergines in spice, cucumbers in spice, tomatoes with herbs, sauté potatoes, green beans in sauce, bread, salty goat cheese, delicious thick cream, several kinds of mushrooms in spices (avidly collected from the woods), and, of course, loads of wine and countless toasts. Then, with the family being of Greek origin (some of its members were working in Greece, too), it was time for music and dance. The daughter, a charming girl whose goal in life seemed to be only to please, put some Bouzouki on their stereo and invited me to dance. Naturally, our dance ended with a round of applause. Then, at 11:00, the party broke up and we went home at last.

Two days later the neighbours all came to bid friendly farewells, and we drove off through the village, past the potholes and the animals. We called in at the Forestry Office where there was a crisis meeting discussing how to manage logging without any diesel fuel because of the shortages. We also visited an ancient sawmill where I saw their antique frame saw which would look good in an industrial museum.

As we were passing by Gori, the birthplace of Stalin, we drove through the town which turned out to be relatively prosperous. I had not expected quite the ostentation of the Stalin place of pilgrimage that I saw. First there was the large square with what I was told was the only statue of Stalin still standing, and then the place of pilgrimage itself. There was a long, tree-lined avenue with a large museum at the end, built in the most ostentatious Russian style. Beside it was Stalin's own railway coach; in front was his very small house with something akin to a mausoleum built over it. Despite the show of grandeur, however, the museum and railway carriage were closed and totally deserted. There was talk at that time of pulling down the museum building, but that seemed regrettable to me because, even if Stalin was hated, the future would want to see the Rogue's Gallery.

Back in Tbilisi, a day was planned for gentle progress to the house of Misha, Giorgio's nephew, in a village deep in eastern Georgia. But first we paid another visit to Rustaveli Avenue where I hoped to find some presents to take home for the family. We discovered a small shop allegedly selling souvenirs, but it had only the leftovers from a vanished tourist trade. I managed to find one Georgian goblet which was nice but crazily cheap. Then we visited the shop which the guidebook said was "the Shop" to visit in Tbilisi. On the outside it said "International Shopping" and looked quite promising, but on entry we found that it was probably the worst victim of the unstable situation and was almost completely empty. "No tourists, no souvenirs," they said. There was nothing Georgian and only one or two Russian scarves and a few shoes.

We moved on, this time to visit the Metechi Palace Hotel and the transformation was almost unreal. A large Hilton-type of affair, it had been built recently by Austrians. We turned up as a couple of tramps and walked through the front door, ushered in by uniformed flunkeys and through an airport-style X-ray screening arch! This was the first time I had ever walked into a hotel like that. Inside it was incongruously ostentatious. While outside there was poverty, huge fuel queues, ruins, and uncertainty, inside it was plush in the extreme. There were two shops within the hotel, of the kind you would expect in five star hotels, full of perfume, whiskey, etc., and not a thing anywhere near Georgian.

Next Misha drove us off into the backstreets of Tbilisi to find his friend who was a popular singer in Georgia. Misha's driving was effective but ignored most of the rules of the road. We found his friend Teimo (Teimuraz Revazishvili to be exact) and then drove to the bullet-scarred Television Centre. They left me while they disappeared within, to agree upon an appearance contract for Teimo and his group of singers.

At length we got underway and headed for the Kakheti region in the east where Misha lived. It was about a 150 miles away, but it was slow travelling because of the hindrances – the car which "pinked" on the poor fuel, potholes in the road, animals suicidally ambling across the road, and people following their example. There were people selling meat placed on plastic sheeting by the roadside as well as all the other things being hawked.

We finally reached Misha's village and called into several places – wine factory, collective farm, vineyards, and two houses. Then at last we reached Misha's house which was quite impressive but lacking a lavatory. There was only the most ghastly hole in the ground in a shed with no door. This really was a croucher of the worst kind.

We were joined by several friends and given a good meal. Teimo had his guitar and was persuaded to sing which he did beautifully. They were wistful songs and two of the men harmonised to the tunes. Then we went to bed in a sort of mini-dormitory with four beds.

The next morning I woke up to the sound of snoring from my fellows in the dormitory. After a moderate breakfast we headed off to another village with wonderful views of the Caucasus Mountains. We drove to Giorgio's third house and met the village carpenter who still had a piece of Pitsunda pine wood from material Giorgio had given him to make new doors for his house. Pitsunda pine is restricted in its distribution and so this was exciting. The carpenter gave me a lot of help getting this sample and planing it for me. Giorgio added other woods, making it a good day's haul, and then we picked up Teimo and found our way fitfully back to Tbilisi. Here Teimo invited us into his house and gave me a record of his singing. He also gave us wonderful hospitality before we returned to Giorgio's apartment. The television was reporting on a serious situation in western Georgia, with a very severe and tired-looking Eduard Shevardnadze talking for a very long time. Apparently it was very bad.

It was now only one day before I was due to head back home, and the drama began to unfold. On Sunday, October 4, I went with Giorgio to the Aeroflot office in Tbilisi. They seemed reasonably confident about the flight the next day.

Back at home, Giorgio's friend, Arsen Pochkhua, came to lunch. Arsen was famous throughout Georgia and worldwide for his wonderful carvings in Caucasian boxwood. The tree is larger than our English one, and he used the marble-like cream-coloured texture to create sensuous faces of exquisite charm. Interestingly, he also cut crystal and was going to see President Shevardnadze to present proposals for a large ornamental crystal vase that Shevardnadze would present to President Bush as a farewell present. As for me, I was fortunate to buy one of Arsen's superb boxwood carvings of the face of a lovely Georgian damsel, which has become one of my treasured possessions.

The next day (Monday, October 5) we awoke early and Giorgio kindly took me to the airport. We arrived at 7:00 am, as was required for the flight at 8:40 am, but there was no indication the flight would be there. Of course this was because of the desperate fuel situation. Not surprisingly, I was the only person at the Intourist (foreign nationals) departure hall.

I said a last thank you and farewell to Giorgio, and I gave him my camera which he admired so much. I was very pleased to do this as he consistently refused to let me pay for anything.

Finally we parted and I was on my own. I began waiting patiently for my flight, hoping I would reach Moscow in time for the connection to London. At 10:30 am I was told there may be a flight to Moscow at 3:00 pm, but if not there might be one in the evening – or not! Since my flight to London was to leave from Moscow at 4:15 pm, I figured that I'd had it for the day. The woman telling me this was trying to be as helpful as she could be, given all the difficulties, but she also mentioned that she was still trying to get flights today for people who had been trying to depart since September 26!

The terminal building had no facilities except lavatories – no shops, no food, no drinks, no notice boards, destination boards, or announcements. Most of the time I was completely alone, but from time to time other people wandered in and out, and staff and scruffy armed policemen strolled past. Eventually, I was joined by an Algerian and an Indian student who were both studying at Tbilisi University on grants. The Algerian spoke quite good English, but conversation with the Indian presented a problem because his strong accent made him almost unintelligible.

Finally around 2:00 pm they indicated that a flight was imminent – well, sometime soon. I checked my baggage, which was grossly overweight with my wood samples. Ironically, this was the first time that I had to pay extra.

At last the long fitful process of getting on the plane began. Tickets were inspected. Gradually a crowd developed in the departure area. Near the door to the tarmac, a scrum (rowdy crowd) built up, but then they decided to move us into another room before allowing us onto the plane. Unfortunately, they couldn't open that door, so the scrum grew. They used a large screwdriver to wrench open the door, but this led to little further progress because we were not allowed any further. Then, some of us were herded into a minibus and driven to the plane while others walked. At the steps to the plane, we were stopped for some considerable time, then at last allowed on slowly. By now, the scrum at the bottom of the steps became intolerable. Then, incredibly, they ordered the people off the plane again! The crowd waiting to come up, however, refused to move back, so I was caught in a frightening crush. Fortunately, that didn't last long, and then we were back fighting our way onto the plane again. Yet with everyone on board, we still could not go and there was a further interminable delay.

At long, long last, at 3:40 pm, seven hours late, we took off to my great relief. The plane was archaic, 30 years old I was told, and smelly and filthy. There was no food or drink on the 2-1/2 hour flight, and nothing to read, but the flight was uneventful and satisfactory.

On arrival at Vnukovo Airport, I was far too late to get my connection to London, the souvenir shop where I had planned to get some Russian dolls for the family was closed, and the taxi service to the International Airport, an hour's drive away, was also closed. I decided it was wise to travel to the Sheremetyevo Airport, since then I could see about getting a morning flight to London. I rang Sasha, my travel agent, whom fortunately I was able to contact! He sent a man in a very smart, large car who brought me to the Novotel at Sheremetyevo, driving me through the centre of Moscow which, in the dark, looked very impressive.

Although it cost me £100, I decided to stay the night at the Novotel. I simply could not face any more hassle or trying to sleep on a bench. Also, I was hampered by my inability to move far because of the fiendishly heavy case, weighed down not only with samples but also with last-minute presents from my hosts. At the Novotel, Sasha came in to meet me and we had a pleasant chat. Suddenly, I was back to normality. The Novotel was excellent, highly civilized, and I enjoyed my first "normal" meal for days and my first bath for two weeks. Getting a copy of the *Moscow Times*, an English language newspaper, I read the headlines of very nasty developments in the Abkhazian region of Georgia. I'd had fun and enjoyed marvellous hospitality, but I was really relieved to be out.

After a pleasant relaxed breakfast the following day, I wandered onto the terminal 400 yards away where I could book and pay for a new flight. With plenty of free time before the afternoon flight, I had a light lunch at the Novotel and then collected the luggage to take it to the terminal. Going through the check-in procedure and getting through customs was no trouble, although my adrenalin was rising at the thought of clearing the last hurdle. Of course, I then hit a problem – a could not get my boarding pass because my visa had expired one day ago. The Russian girl working for

British Airways offered to put my case on one side while I went to emigration to see if they would clear me through. After some delay and muttering, they finally let me through, but then I wanted to go back and put my very heavy case (far too heavy for her to lift) back onto the conveyor and into the system again. I asked to go back, but the emigration authorities flatly refused to allow me.

I was now extremely worried, thinking I had effectively lost the case with its vital samples. Verbal communication was difficult because of the language barrier, but it went something like this:

> "You have now emigrated; it is not possible to go back."
>
> "But I only want to ask that girl to put my case back on the conveyor, and it is so heavy that she cannot possibly lift it."
>
> "Nyet."
>
> "Can you please tell her I have emigrated and help her put the case on the conveyor for me?"
>
> "Nyet."

Panic.

Then a real conversation with a suave person who materialised from nowhere:

> "Can I help you?"
>
> "If you can speak Russian you certainly can."

I explained the problem.

> "That's not a problem at all. I am a British Diplomat here. I can im migrate and emigrate all day long with this pass. Give me your boarding card."

He strolled like magic through the impenetrable barriers (and I wondered if I would ever see my boarding card again). Mercifully he returned.

> "Good God, you must have half of Russia in that case. Here's your boarding card back."

Thank God for British Diplomats.

Just one more Russian obstacle then stood in the way – the gangplank. The flight was called, I showed my treasured boarding card, and, stepping along the gangplank toward neatly groomed Western civilisation, with a deep sigh, I climbed aboard.

Chapter 19
The Wander Lust

I thought at the time that my Georgia trip had cured me of my wander lust and that I would never want to risk experiencing anything like that again. But, of course, it was really a balance of exhilaration and desperation. It wasn't too long before exhilaration tipped the balance again.

The next year, 1993, after all of that wood hunting, I decided I would indulge my other interest: history. This time, wood would not be the focus for I was going to Egypt to treat myself to a feast of ancient remains.

But as always, it didn't work out like that.

If there were no River Nile, there would be no Egypt. It may be over 4,000 miles long, but the Nile is not wide, varying from a quarter to a half mile at most. Along its sides are relatively narrow strips of rich alluvial, well-watered soil. Just about every square mile of that is cultivated intensively. Outside that is mile upon mile of completely barren desert with virtually no rain. So how can a wood nut return from Egypt with his brain reeling about trees and wood?

Well, he should start at the Egyptian Museum of Antiquities in Cairo (and he could easily finish there for it has 100,000 exhibits) and see the Treasury of Tutankhamen. Nearly the complete contents of 400 treasures of King Tut's tomb are on display, and the effect is mind boggling. It takes an act of humility to adjust to the fact that the quality of workmanship is at least equal to the finest anyone could produce now, and a gasp of unbelief that the condition is such that it looks a mere 100 years old, not the more than 3,000 that it is.

Most of the exhibits are contained within glass cases for protection from people, dust, and varying humidity. One can still get close enough to see that the most prized wood was black and said to be ebony, but I have a feeling that much was actually African blackwood. Their ancient stories are of voyages to the Land of Punt, which is thought to be Somalia and reliable supplies of blackwood seem more likely than ebony. (Likewise, in a bazaar I bought an antelope carving from Sudan which was sold as ebony but is definitely blackwood.) In the exhibits, the finest pieces are of the blackwood inlaid with ivory and stained woods. The quality of intricate carving and inlay is imaginative, meticulous, and superb.

Of course gold was the greatest treasure, and it is there in abundance. It was either solid or it was a wafer-thin overlay of gilding on wood as in King Tut's series of four shrines fitting inside each other. The only readily available native wood was the sycamore fig, which is soft, coarse-textured, and, in these dry conditions, reasonably durable. It could be sanded to a smooth enough finish, and it was the favourite wood for coffins

The most widely used semi-precious wood was Cedar of Lebanon which fitted the needs of the ancient Egyptians quite well. It was available in large sizes with relatively few defects, it was quite easy to work, and it smelled delicious. They built thrones, chairs, beds, cabinets, and boxes of staggering complexity and workmanship. Some of these made me despair – why can't I do work half as good as that?

The 4,700-year-old boat to take the Pharaoh Cheops to the Next World (found buried and dismantled next to the Great Pyramid in Cairo in 1954 and reassembled).

But Lebanon is across the sea. To get the cedar, they not only had to devise how to navigate the water but also how to build boats to make the trip. For this they built marvellous boats of cedar.

By the Great Pyramid of Cheops on Cairo's outskirts, a dismantled boat for reassembly was found in 1954 in a sort of cellar. Made for Cheops' journey to the next world, it wasn't used for that. It had been placed there around 2690 BC or about 4,700 years ago. Taking 15 years to complete the reassembly, it now stands at a remarkable 140 feet long, 18 feet wide, and is in a wonderful state of preservation.

This boat was powered by oars, but in Cairo Museum, contemporary models not much newer have fully rigged lateen sails which could be raised and lowered by pulleys.

The ancients were highly organised timber importers, and wood availability was not an insurmountable problem. I was fascinated by how they worked the wood. In the dry climate, there are remarkable survivals, not only of the tools themselves as in the Cairo Museum but pictures of the tools in use.

It is uncanny to enter the many tombs dating back 3,000 to 4,000 years and see wall paintings of exquisite charm and skill covering every available surface, looking so fresh that they seem to have been done yesterday. In the tombs of the kings, these portrayed the king's magnificence; in those of the nobles and senior craftsmen, they show how the end results of woodworking and other skills were achieved by slightly different means from those of today. In Cairo Museum is a superb, small wood model, 4,000 years old, of 15 carpenters working with various tools in a workshop, which gives more clues as to how it was done.

The Egyptians had copper but not iron, so with this slightly softer metal they could work the softer woods such as cedar and sycamore fig. Their saw was very similar to ours with a long handle in a straight line with a copper blade. The means of securing the wood was an ingenious alternative to our vices, clamps, and benches. A round pole squared on one edge was fixed vertically in the rocky ground and the piece to be sawn was

Model of an Egyptian carpenter's workshop with men at work, about 2000 BC.

strapped vertically to it. In some cases, a stabilising weight was used in the form of a heavy stone slotted through the straps so as to keep the wood really tight. So that the saw would not jam, a blocking piece was wedged in at the top of the wood to be sawn.

They had not discovered the plane (the Romans taught them that much later), but they used copper adzes, which are like mattocks but ground very sharp and used along or across the grain. They were short-handled, only about eight inches. The Cairo Museum shows a team of four adzing a plank. It seems that a relatively fine surface could be obtained. In place of sandpaper, they used blocks of sandstone, and no one would be able to improve on the surface they achieved.

Their chisels were indistinguishable from ours, and their mallet heads were rounded as used by stone masons and wood carvers today. With these they made tight mortice and tenon joints and carved to very fine tolerances.

I found the drills most fascinating of all. They could be operated by one or usually two people. The copper drill was made so that it cut in either rotating direction. This drill with a wood head was held down with pressure by the first man while a second man with a bow had threaded the bow-string once around the drill. As he moved the bow backwards and forwards, he rotated the drill.

Egyptian craftsmen at work including, in the middle row left to right, woodworkers drilling, sawing, adzing, sanding, and painting. From the tomb of Rekmire, about 1440 BC.

For measurement they had rulers in cubits, and they had squares, plumb lines, and water levels.

Wall paintings show craftsmen painting and varnishing in much the same way as we do today.

The end result is mind blowing.

Nowadays Egypt imports softwoods and hardwoods worldwide. I saw no plantations for timber in my travels over much of the fertile Nile side; the space is too valuable for raising food crops. The street trees of Egypt, especially in Cairo, however, are a tree-lover's paradise. Everywhere are the Flamboyant trees with their striking red flowers and two-foot-long pods, jacarandas, cassias fat bottle trees, figs of various species, royal palms, date palms, and many others, all making a very pretty sight.

~ ~ ~ ~ ~ ~ ~ ~ ~ ~ ~ ~

In 1995, I visited Peru. Again my reasons for going were not really focused on wood. I went to see the spectacular sights of Machu Picchu and the old Inca remains around the town of Cusco, where a stonemason would be as overwhelmed by the workmanship as a woodworker seeing ancient Egypt.

There were interesting examples of the effect on vegetation of climate and altitude. On the coast facing the Pacific, there is virtually no rainfall and these relatively flat lands are totally barren desert. That is except where the rivers come down westwards from the Andes. Here, as in Egypt, the land by the rivers is fertile and green, and cities such as Lima flourish, helped by a damp mist when it drifts in from the ocean. As you climb up the west side of the Andes, vegetation begins and the steep and lofty mountains, like huge canine teeth, are covered with plentiful low trees. The east side of the Andes (which I did not see) are covered with lush vegetation. Here would be a wood hunter's paradise. But even in my short visit, with a little local help, I did not come home empty-handed.

I paid a similar visit to India in 2001. Thanks to a great deal of help I have had over many years from research centres in India, I did not expect to add to my collection. Instead, like so many others, I marvelled at the beauty of the Taj Mahal. In its gardens is a wide range of trees, so I could see from which kind of tree my wood specimens had come, such as banyan, frangipani, sissoo, rosewood, and asok.

It was fascinating, too, to see for sale in the shops within hotels, examples of the highly skilled woodwork for which India is well-known. There

The Great Cypress of El Tule, Mexico, thought to have the world's largest diameter trunk at 46 feet.

was small furniture, exquisitely carved boxes, and figures of gods and of animals. I was particularly intrigued by a small figure of an elephant carved in sweet-smelling sandalwood. The sides were drilled, carved, and hollowed out like a lace covering, revealing the figure of another elephant standing totally within the first one, all made out of one piece.

A visit to Mexico followed the next year. Again my purpose was to see history, but again it was a fascinating place to see the lush forests and beautiful trees. One tree I had to see was the Great Cypress of El Tule. This is by no means the tallest tree in the world, nor the oldest, nor the biggest, but it lays claim to having the largest circumference of trunk in the world at 190 feet, or 46 feet in diameter. There were also beautiful street trees to be seen such as the Flamboyant trees in the city of Oaxaca.

A visit to China rounded out the "series." Again it was aimed at history rather than trees or wood, but there was still a chance to see some of the huge number of native tree species of the country.

Chapter 20
Wild West – Again

During all this time, of course, the collecting urge continued as ever. I have very good friends in Los Angeles, California and Boulder, Colorado.

I paid three visits which took me from San Francisco, California to Texas. I drove down the coast road from San Francisco to Los Angeles, taking in the beauty of Big Sur and Monterey with its pines and cypresses. My friends in Los Angeles showed me all of the best sights, including Hollywood, Bel Air, and Malibu. I twice took my chance in Las Vegas and later in sweltering Death Valley, where the temperature was 112°F, still by no means a record.

I was very keen to see the remains of ancient Indian settlements, taking in Mesa Verde with its houses sheltered under a massive natural roof, Canyon de Chelly with the houses tucked into sheer cliff sides, the ancient village of Acoma, in New Mexico west of Albuquerque, perched on an isolated "island" mesa, as well as Chaco Canyon, the remains of an Indian "city" now "in the middle of nowhere" in New Mexico. There was a climb to the 14,000 foot top of Mount Evans in Colorado (almost all the way up by car), as well as many other sights, including Monument Valley of "Stagecoach" fame and the Arches National Park, all of which I thoroughly enjoyed.

But it was while I headed south to southern Arizona that the heat built up (which I like, but some hate!), and the scenery began to change to what I love – the saguaro, organ-pipe, and jumping cholla cacti, creosote bushes, mesquites, and paloverdes, as well as another visit to the OK Corral at Tombstone and a day south of the border, down Mexico way, at Nogales.

While the objective of the visit was seeing these sights, it was not without its rewards in collecting yet more samples with the help of friendly locals.

Meanwhile, my very good friends of my first visit to the southwest, Alan and Mary Ann Curtis of Oregon, asked me if I would like to come to a meeting of the International Wood Collectors Society in Tacoma, Wash-

Craftwork on display at the International Wood Collectors Society meeting in Tacoma, Washington, 2003.

Interior of the mass-production, computerised sawmill.

ington. I had never been to an international meeting of the Society before and so I readily said yes. It was a wonderful experience. The Society had just about taken over the Sheraton Hotel and there were displays of superb workmanship, and stalls of wood, tools, craftwork, and books for sale. There were talks, sawing of wood from logs, an opportunity to exchange wood specimens, and a tempting wood auction. We also had the opportunity to visit a massive sawmill which was producing out huge volumes of lumber with just a handful of people controlling the computerised operation.

All of this was good, but the best part was meeting people with whom I had corresponded for years or who had been just names to me. The atmosphere was wonderfully enthusiastic and friendly.

Some of the coast redwood (Sequoia) trees in northern California, among the tallest in the world, which reach about 350 feet.

Looking up into some of the tallest trees.

After this, Alan and Mary Ann invited me to their home in Oregon, where there was no doubt that Alan's enthusiasm for wood was just as passionate as mine.

We also went down to the northwest tip of California to see what are said to be the tallest trees in the world – the coast redwoods. We called in at the Visitors Centers to seek to find the world's tallest, but each center seemed to be championing a different one. It is pretty difficult to measure such trees to an exact height. What seems to be fairly certain is that there are several in the area reaching at least 350 feet, and we hope we saw some of those. One tree that does not reach quite that height but is easy to reach and great to see is "The Big Tree" by the road near Orick in the Redwood National Park; it stands at 304 feet tall. While these trees vie to be the tallest, they are certainly not the world's biggest nor do they have the diameter of the Great Cypress at El Tule, Mexico, but they are a great sight.

So, at the end of that visit, I now had not only a wonderful time but might, at one time or another, have seen the tallest, biggest, greatest diameter, and oldest trees in the world. What a way to finish my wanderings.

Chapter 21
Viva Argentina

But, of course, it didn't end there! By 2007, I had established contact with International Wood Collectors Society members Alex and Maria Nencini in Buenos Aires, Argentina. Alex, a wood craftsman, had helped me with several woods from that vast country. Then, by coincidence, our daughter and her husband had moved into a house fairly near to us and were having a housewarming party. Who should we meet there but another Alex and Maria from Buenos Aires. This Alex is a doctor now living and working in England. I talked to them both and marvelled at the coincidence when Maria said, "You should go over to Argentina and meet my parents – my father is a carpenter."

Well, of course I did. The "other" Alex and Maria were very enthusiastic about this and the prospect of our teaming up to look for some more sample material in Argentina. I felt that Alan Curtis, my co-explorer, would enjoy this too.

To cut a long story short, we arranged that we would stay in Buenos Aires for a few days to see the sights. Alan arrived a day before me, having stopped in Brazil along the way. Alex very kindly picked me up at the airport in the middle of the night and took me to the hotel where we were to stay. On our arrival there seemed to be a horrible smell in the bedroom that Alan and I were sharing. It turned out to be some material of a revolting-smelling wood Alan had gotten in Brazil and which we soon dubbed "El Stinko." So strong was the smell, and so late in the night were the new arrivals from different parts of the world, that the hotel thought we were drug dealers. Alex kindly took El Stinko off to a place outside his home to calm down the situation.

We had a great time looking around the pulsating city of Buenos Aires with its 13 million inhabitants – a third of the population of Argentina. Alan and I blundered our way around the city getting lost several times and hiring taxis to rescue us. Although Argentina had recently been through a bad economic patch, things were looking much better and the up-market centre of the city was looking buoyant. We visited the Plaza de Mayo and the shopping street of Calle Florida, as well as the prosperous district of La Recoleta with the huge cemetery where Evita (Eva Peron) is buried.

We looked at the large number of bookshops and found good books about Argentina's trees, and then went to the Botanic Garden. Until recently it had been sadly neglected, but now it is looking very good. Of course we gravitated to the wood pile and were not disappointed.

At Alex and Maria's house, Alex had already gone to a great deal of trouble assembling some very interesting material for specimens from nearby and from two separate, distant areas of Argentina, the southwest and the northwest. Again the collection was growing.

Our main objective was to visit the most wood-productive area of Argentina in the northeast province of Misiones. Misiones resembles a finger pointing upwards at the extreme northeast of Argentina. This is surrounded by Brazil, Paraguay, and Uruguay. It is also the home of the mighty Iguazu Falls; these are perhaps the world's greatest group of waterfalls, but they are very little known because of their remoteness.

Palo Borracho trees in the Botanic Gardens, Buenos Aires.

We set off in our hired car to begin the 800 mile journey to Iguazu. Soon after leaving Buenos Aires, we entered the classic pampas and gaucho country that one expects in Argentina – miles and miles of flat pampas with estanchia (ranch) houses spread at great distances apart. In groups were huge numbers of cattle, sometimes with gauchos who were always in pairs on horseback herding the cattle.

As we headed north, we entered vast plantations of eucalyptus and pine and large sawmills. Here the climate is right as it is warm and quite often wet, and it even rained torrentially for one full day as we drove up.

The Iguazu Falls on the border between Brazil and Argentina.

After two days traveling, we had lunch with a very nice family of German origin who had a large sawmill and forestry land. They had volunteered to help us with sample material of species new to us, and they would have this ready for us in three days upon our return from the Iguazu Falls.

It is a pity that the Falls are so remote, because they really are a spectacular sight. They are arranged on a horseshoe-shaped precipice about two miles around, on the border between Argentina and Brazil. The many falls are best seen by walking along trails through the sub-tropical rain forest. We spent a wonderful day gazing spellbound by it all. On the following day, we took a taxi over to the Brazil side of the falls and marvelled again.

On our way back, after looking at the remnant forest of parana pine trees which are now being conserved after near oblivion, we came back to our hosts. They had collected over 20 species for us from their own land and insisted that they did not want to be paid.

After leaving them, we took a diversion westerly, seeing more of Argentina, and returned to Buenos Aires having covered 2,000 miles.

A few days later, Alan and I thanked Alex and Maria for all they had done for us and we returned to our respective homes with some valuable additions to our collections.

Chapter 22
What of the Future?

So, safely back home, my collection had now passed 7,000 species – a great excitement and relief as it had been quite difficult reaching that target and it was nice to relax.

I've said it before, but I'll say it again – I think now my collecting will slow down to a snail's pace. As I write this, I have just reached 7,100 and by coincidence I am 71 years old. Several factors affect my thinking. Although the world has many times the 7,100 species in my collection, there tends to be a multiplication of species in one genus (for example, about 400 species of oak) with very little difference in the wood, so probably all the "must get" woods are in my collection. Then there is the complex world of synonyms. It is said in the botanical world that the subject is peopled by lumpers and splitters. One botanist declares that a few species are actually one and writes a learned thesis about it, lumping them together. A few years later, along comes another botanist who says that another species is actually several and writes another thesis that splits them apart. The effect of this is to create a minefield for a collector – you get new woods only to find that, despite your best efforts, you have duplicated something. Yet another factor is space. My storage and display system seems quite effective in that wood specimens can be seen side-by-side, but it takes up a lot of room. There is a tendency, too, to think that one has seen every kind of wood there is, a bit like a "been there, seen that, got the t-shirt" affair – but something new crops up and off we go again.

People ask me what will happen to my collection one day. I am pleased to be able to say that the Royal Botanic Gardens, a World Heritage Site at Kew in London, which has a world-renowned research facility in the Joddrell Laboratory, has inspected my collection and is keen that one day it should be passed to them. I am pleased that it will be.

But I really sincerely hope to be able to add just a few more specimens to it before then!